A Journey into Gravity and Spacetime

A JOURNEY INTO GRAVITY AND SPACETIME

John Archibald Wheeler

SCIENTIFIC
AMERICAN
LIBRARY

A Division of HPHLP
New York

Library of Congress Cataloging-in-Publication Data

Wheeler, John Archibald, 1911–
    A journey into gravity and spacetime / John Archibald Wheeler.
        p.   cm.
    ISBN 0-7167-5016-3
    1. Gravity.   I. Title.
QB334.W49   1990   89-24256
531′.14—dc20   CIP

ISSN 1040-3213

Printed in the United States of America

Scientific American Library
A division of HPHLP
New York

Distributed by W. H. Freeman and Company
41 Madison Avenue, New York, New York 10010
20 Beaumont Street, Oxford OX1 2NQ, England

1 2 3 4 5 6 7 8 9 0   KP   9 9 8 7 6 5 4 3 2 1 0

This book is number 31 of a series.

Truth is less than truth until
it is made known.

Contents

Many a fair young friend, oh Gravity,
By smile and happy word
Has made my heart beat faster,
But with you my love affair
Has never ended.
You grow ever more beautiful
With each passing decade.
You lead the way
To a new and higher lookout point,
And behind yet another mystery
You reveal hitherto hidden simplicity.

Who first held up to me the mirror,
Oh happy Gravity,
Guardian of so much wisdom,
That gave me a first glimpse of your charms?

Shall I thank most J. Arthur Thompson
Speaking through his
Great Outline of Science
To a boy of ten, fresh indoors
From feeding the pigs
And herding in the cows?

Or are the thanks still warmer
From the boy of thirteen
To his friend
For the loan of that little book by
Charles Proteus Steinmetz,
Four Lectures on Relativity and Space?

Fifteen I was, oh Gravity,
When you first directly cast on me the spell
Of your attractions and your mystery,
Not through any mirror or reflected image,
But by your own alluring voice.

A swing I had often seen
Carrying its passenger in ever-higher arcs
By reason of a succession
Of modest but well-timed pushes.
That magnification of a small effect
By resonance into a big effect
Made it easy to grasp the central idea
Of the experimental measure
Of the attraction of mass by mass
Long ago dreamed up by John Michell
And tried by Henry Cavendish.

Now before my eyes
Paul Heyl, busy but calm and warmly welcoming,
Master scientist of the National Bureau of Standards,
Was performing that very experiment.

Weak though you are, oh Gravity,
In drawing together two everyday masses,
I saw with my own eyes
This weak pull magnified up by swinglike resonance
To an effect which Heyl was then and there measuring
To a precision up to that time unprecedented
And destined to stand for many years.

Smiling at my fascination with your force, Gravity,
Heyl took me to another room
And showed me the machine
He had built of gears and pulleys
And a string running over those pulleys.
The end of the string grasped a pen.
The pen traced out on moving paper
An hour-by-hour prediction of the height of the tide.
That machine I wish I had here now
As I look down from my window
On the ten-foot tides of the Atlantic,
Raised by the force you carry, oh Gravity,
From Sun and Moon.

Floating over the open sea
I see a white cloud
Forming, one by one, as the moments drift by,
The face of many a friend
And colleague of later days
From eager-eyed questioning freshman
To deep-digging graduate student
And from co-author of book or paper
To stooped white-haired venerable source of wisdom,
All inspirations, Gravity,
In the study of your beauty.

Help me to make this account
A radiant testimony
To the wonderful simplicity
Of the principle that the boundary of a boundary is zero,
Heart of Einstein's great 1915,
Battle-tested, and still standard
Geometric account of your action, oh Gravity.

J. A. Wheeler

Nothing more strongly animates this book than a passionate belief that this is *our* universe, our museum of wonder and beauty, our cathedral. Marvelous are the seamounts and the geysers of the ocean floor. Dream-provoking are the buried cities of the past. Awe-inspiring are the continental drift that steadily tears Iceland apart and the fiery furnaces that we call stars. The pace of exploration grows swifter with each passing decade. We celebrate each new finding, each new glory, each new wonder. They are the rightful inheritance of every thinking human.

The wonders found take second place to those still waiting to be found. The undiscovered lies around us in every direction, at what the ancient Romans liked to call "the flaming ramparts of the world," *flammanda moena mundi*. How humbling it is to realize that we, today, knowing so little about the world and ourselves, live still in the childhood of mankind. Yet what a joy! What a challenge! It is not surprising that the greeting between eager searchers of the unknown is often not "Hello," not "How are you," but "What's new?"

Each year brings new findings about the mechanism of inheritance, the structure of matter, the workings of the brain, and many another active frontier. How happy it is for the interested inquirer from the larger community when the new result links familiar, everyday observations with some deep insight into the machinery of existence. Gravity, among all domains of investigation, makes the tightest connection—and the tie easiest to follow—between the simple and the profound, between the old and the new, between what we see around us and the inmost workings of the universe.

Einstein's battle-tested general theory of relativity tells us that gravity is not a foreign and physical force transmitted through space. It is instead a manifestation of the curvature of that four-dimensional union of space and time that we call spacetime.

Gravity is more understandable than any other force of nature. We gain our deepest, surest insight into gravity today by sticking to the familiar world of space and time. Naturally, we have to pay the price. We have to learn—but will find this not too hard—to think of space and time as unified into spacetime, and to think of this spacetime as curved.

How is this abstract world of curved spacetime geometry wired up to the everyday world of tennis balls and falling weights, of spaceships and planets, of stars and galaxies? The answer is simple yet wonderful: spacetime geometry is wired up to the everyday world by a geometric principle of fantastic innocence and power, a principle that says that "the boundary of a boundary is zero." This boundary principle reaches out guiding hands from every region to the surroundings of that region. In this way *spacetime grips mass, telling it how to move*. In this way *mass grips spacetime, telling it how to curve*. With those two statements in hand, we

Preface

hold before us in a nutshell all of Einstein's great geometric theory of gravity.

Out of a view of gravity so abstract springs our understanding of happenings as concrete as the fall of a stone and the float of an astronaut; the rhythm of the tides and the stretch of the oil prospector's gravimeter; the jiggle of a gravity-wave detector and the course of a comet; the bending of light by the Sun and the slowing of the expansion of the universe; the collapse of an overdense star to a black hole; and the generation of power by that fantastically massive black hole that is now spinning away at the center of our own galaxy of stars, the Milky Way.

All I hope to explain here about these miracles of nature I owe to the writings of the thoughtful, to discourse with colleagues in and out of academia, and to being taught over the years by those I taught, by their puzzlements and smiles, by their researches and papers, by their ex-postulations and enthusiasms: young people, majoring in English or in physics, in business or in chemistry, in art or in political science, in law or in architecture. It makes my writing here a joyful undertaking to think of them, of what they have taught me, and of what they would want me to say.

John Archibald Wheeler
Princeton University, 1989

A Journey into Gravity and Spacetime

1

Great Men, Great Ideas

Oh great thinkers of the past,
You identified and solved
Many a mystery of motion
Where others could see no mystery.
The faraway and strange
You linked to the near and clear—
Fall of apple to orbit of Moon,
Inertia here to influence of mass there,
Speed of light to pattern of spacetime.
Thanks to your hard-won insights,
The whole great story of gravity
Now comes to us
In a single simple sentence:
Spacetime grips mass,
Telling it how to move;
And mass grips spacetime,
Telling it how to curve.
By your own childlike spirit of inquiry
Remind us that we, children all,
Only little by little are seeing with new eyes
This strange and beautiful universe,
Our home.

Isaac Newton *Born December 25, 1642, Woolsthorpe, Lincolnshire, England.
Died March 20, 1726, Kensington, London.*

No force of nature presents itself to humankind more directly than gravity, none has attracted to its study over the centuries greater thinkers, and none has opened up—and continues to open up—grander perspectives. The action of gravity encompasses the arc of the Moon across the sky, the fall of the brick from the cornice, and the sweep of the Earth around the Sun. It powers the gravity waves that carry energy away when tremendous mass clashes against tremendous mass. It crunches space in the interior of a black hole, bringing a time after which there is no time. Even more inspiring are the billions upon billions of cubic lightyears of space added to the volume of the universe every second, explained by gravity's action.

In the new and larger view we have today of gravity, the innocent emptiness that fills the heavens turns out not to be innocent. Space and time, unified as spacetime, do not merely witness great masses struggling to bend the motion of other masses. Like the gods of ancient Greece, spacetime helps guide the battle and itself participates. The doorway to the workings of gravity was opened early in this century by Albert Einstein in his geometric theory of gravity—"geometric" because its key lies in the geometric shape of spacetime. The scope and power of this century's new view of gravity and spacetime is seen nowhere more dramatically than in its prediction of the expansion of the universe. To have predicted, and predicted successfully, and predicted against all expectation, a phenomenon so fantastic is the greatest token yet of our power to understand this strange and beautiful universe.

Newton's Ideas on Space and Time

Isaac Newton, in his great *Principia,* first formalized our understanding of gravity as a force that acts at a distance through space, drawing any two masses together. It is often thought that Einstein's geometric theory of gravity proved Newton's concept of gravity to be wrong. But Einstein never spoke to me of anything more often or more forcefully than his admiration for the courage and the judgment of Newton.

Newton knew how difficult it is to describe motion. We cannot perceive motion in itself; we can perceive motion only relative to some other object. Do we then measure the motion of a bullet relative to that waving branch, or to that galloping horse, or to those faraway stars? Newton had the sense to see that the question was too difficult to answer in his day. He therefore assumed the existence of "absolute space," an invisible empty space that is at rest relative to any motion at all, and he proposed that all motion be measured relative to this absolute space.

I do not know what I may appear to the world, but to myself I seem to have been only a boy playing on the seashore, and diverting myself in now and then finding a smoother pebble or a prettier shell than ordinary, whilst the great ocean of truth lay all undiscovered before me.

As quoted in Brewster's Memoirs of Newton

That one body may act upon another at a distance through a vacuum, without the mediation of any thing else, by and through which their action and force may be conveyed from one to another, is to me so great an absurdity, that I believe no man, who has in philosophical matters a competent faculty of thinking, can ever fall into it.

Letter From Newton, 1692–1693

Chapter One

Einstein emphasized to me that Newton understood the problems inherent in the concept of absolute space. Nor had the difficulties escaped Newton's great contemporary Gottfried Wilhelm von Leibniz, who declared that "space and time are orders of things, and not things." By "orders of things" he meant that space and time are ideas that we ourselves create in order to put into convenient arrangement the sensory impressions that come to hand and eye. Einstein was to echo Leibniz, saying, "Time and space are modes by which we think, and not conditions in which we live."

Newton had courage, Einstein explained, in that he knew better than his followers that the concept of absolute space could not be defended; nonetheless, he realized that he had to assume the existence of absolute space in order to understand motion and gravity. How else—in those days—could he have hoped to talk meaningfully about gravity? How else would he have been able to conclude in 1666, at the age of 23, that "having thereby compared the force requisite to keep the Moon in her orb with the force of gravity at the surface of the Earth [I] found them to answer pretty nearly." How else would he have known how to establish the paths of the planets through space and account for the tides? The whole science of mechanics became a science only when, paradoxically, it accepted ideas indefensible except through their results:

> Absolute space, in its own nature, without relation to anything external, remains always similar and immovable. . . . Absolute, true and mathematical time, of itself, and from its own nature, flows equably without relation to anything external.

Newton's ideas of space and time, productive but puzzling, goaded Einstein to rethink gravity. New ideas in geometry gave Einstein his first inspiration as to what a new vision of gravity could be.

The Birth of Non-Euclidean Geometry

Not forgotten from high-school geometry is the axiom "parallel lines never meet." More formally, as Euclid (active around 300 B.C.) stated, given any straight line and any point not on that line, we can draw through that point one and only one straight line parallel to the original line. This is the parallel postulate. In the early 1800s the discussion grew lively about whether the properties of parallel lines as presupposed in Euclidean geometry could be derived from the other postulates and axioms, or whether the parallel postulate had to be assumed independently. More than two thousand years after Euclid, Karl Friedrich Gauss, Janos

Gottfried Wilhelm von Leibniz

Born July 1, 1646, Leipzig, Germany. Died November 14, 1716, Hanover, Germany.

For deriving all from nothing there suffices a single principle.

Leibniz

Although the whole of this life were said to be nothing but a dream and the physical world nothing but a phantasm, I should call this dream or phantasm real enough, if, using reason well, we were never deceived by it.

Leibniz

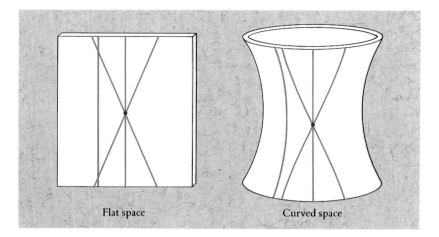

Flat space   Curved space

Bolyai, and Nikolai Ivanovich Lobachevsky independently discovered
geometric systems that satisfy all the axioms and postulates of Euclidean
geometry except the parallel postulate. These geometries showed not
only that the parallel postulate must be assumed in order to obtain Euclidean geometry but, more important, that non-Euclidean geometries
can and do exist.

We do not have to look far for an example of non-Euclidean geometry. In the diagram on this page, we are looking at the two-dimensional
surface of a sphere, itself "embedded"—to use mathematical terminology—in the three-dimensional space geometry of everyday existence.
Euclid's system accurately describes the geometry of ordinary three-
dimensional space (called 3-geometry) but not the geometry on the surface of a sphere (one type of 2-geometry). Thus, the lines AA' and BB'
start going straight north at the equator, and there they are surely parallel. Moreover, granted that they have to stay on the surface of the sphere,
they are as straight as any lines could possibly be. Neither bends in its
course one whit to left or right. Yet they will meet and actually cross at
the North Pole. Clearly, straight lines on the curved surface of a sphere
don't obey Euclid's parallel postulate.

The thoughts of the great mathematician Karl Friedrich Gauss about
curvature stemmed not from theoretical spheres drawn on paper but
from concrete, down-to-earth measurements. Commissioned by the
government in 1827 to make a survey map of the region for miles around
Göttingen, he found that the sum of the angles in his largest survey
triangle was less than 180 degrees. On a flat surface, the three angles of a
triangle add up to 180 degrees. The deviation observed by Gauss—
almost 15 seconds of arc (a quarter of a minute of arc, or $\frac{1}{240}$ of a degree)—

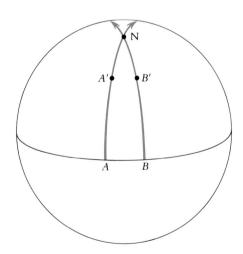

On the curved surface of a sphere, two
straight and originally parallel lines will meet
at the North Pole.

Chapter One

was both inescapable evidence for and a measure of the curvature of the Earth.

To recognize that straight and initially parallel lines on the surface of a sphere can meet was the first step in exploring the idea of a curved space. Second came the discovery of Gauss that we do not need to consider a sphere or other two-dimensional surface to be embedded in a three-dimensional space to define its geometry. It is enough to consider measurements made entirely within that two-dimensional geometry—such, for example, as would be made by an ant forever restricted to live on that surface.

How would the ant know that the sphere is curved? Or, we can equally well ask, how we would know that the Earth is curved if, like our predecessors, we could not leave its surface. The prescription is simple. Take an arbitrary point on the surface and call it N, as in the diagram on the following page. Pick some arbitrary direction and proceed in that direction for 10,000 kilometers, swerving neither to the left nor to the right. At the terminus, drive a stake into the ground. After returning to N, start off in some other direction, go the 10,000 kilometers, and drive another stake into the ground. Repeating this procedure over and over again, we finally plant so many stakes that they describe a closed curve as continuous as we please. If the Earth's surface were flat, then that curve would be a circle with its center at N and with a radius of 10,000 kilometers. By the law of circles, the circumference would be 2π times the radius, or $2\pi \times 10{,}000$ kilometers = 62,832 kilometers. However, because the Earth's surface is curved, the circumference of our staked-out circle does not have this ideal value. Instead, the circumference of the "equatorial" track is four times the distance from the circle to N, or $4 \times 10{,}000$ kilometers = 40,000 kilometers. The discrepancy between this number and the 62,832 kilometers is evidence enough that the surface bends, that the Earth is not flat. Moreover, if we had pursued each straight track from N, not for 10,000 kilometers, but for 19,999 kilometers (1 kilometer short of the point S, antipodal to N), then the circle of stakes would have had a circumference very close to $2\pi \times 1$ kilometer = 6.28 kilometers.

Gauss did not limit himself to thinking of a curved two-dimensional surface floating in a flat three-dimensional universe. In a private 1824 letter to Ferdinand Karl Schweikart, he dared to conceive that space itself is curved: "Indeed I have therefore from time to time in jest expressed the desire that Euclidean geometry would not be correct." The inspiration of these thoughts, dreams, and hopes passed from Gauss to his student, Bernhard Riemann, frail in health, great in mind.

Bernhard Riemann went on to generalize the ideas of Gauss so that they could be used to describe curved spaces of three or more dimen-

Karl Friedrich Gauss

Born April 30, 1777, Brunswick, Germany. Died February 23, 1855, Göttingen, Germany.

Although geometers have given much attention to general investigations of curved surfaces and their results cover a significant portion of the domain of higher geometry, this subject is still so far from being exhausted, that it can well be said that, up to this time, but a small portion of an exceedingly fruitful field has been cultivated.

Gauss, Royal Society of Göttingen, October 8, 1827

The circumference of a staked-out "latitude" line (green) 10,000 kilometers from any point N on the Earth's surface is less than that predicted from the law of circles. Because of this discrepancy, we can conclude that the Earth is curved, not flat, without ever leaving its surface.

sions. Gauss had found that the curvature in the neighborhood of a given point of a specified two-dimensional space geometry is given by a single number. Riemann found that six numbers are needed to describe the curvature of a three-dimensional space at a given point, and that 20 numbers are required for a four-dimensional geometry. There is nothing that requires the curvature at one point to be equal to the curvature at another point, so in general the Riemann measures of curvature generally vary from place to place.

In a famous lecture he gave June 10, 1854, entitled "On the Hypotheses That Lie at the Foundations of Geometry," Riemann emphasized that the truth about space is to be discovered not from perusal of the 2000-year-old books of Euclid but from physical experience. He pointed out that space could be highly irregular at very small distances and yet appear smooth at everyday distances. At very great distances, he also noted, large-scale curvature of space might show up, perhaps even bending the universe into a closed system like a gigantic ball. But as Einstein was later to remark, "Physicists were still far removed from such a way of thinking: space was still, for them, a rigid, homogeneous something, susceptible of no change or conditions. Only the genius of Riemann, solitary and uncomprehended, had already won its way by the middle of the last century to a new conception of space, in which space was deprived of its rigidity, and in which its power to take part in physical events was recognized as possible."

Even as the 39-year-old Riemann lay dying of tuberculosis at Selasca on Lake Maggiore in the summer of 1866, having already achieved his great mathematical description of space curvature, he was working on a unified description of electromagnetism and gravitation. Why then did he not, half a century before Einstein, arrive at a geometric account of gravity? No obstacle in his way was greater than this: he thought only of space and the curvature of space, whereas Einstein, as we will soon learn, discovered that he had to deal with spacetime and the curvature of spacetime.

The name of Ernst Mach (1838–1916) also must figure in any just account of the intellectual antecedents of Einstein's great discoveries. Mach, who was influenced by the earlier teachings of Leibniz, argued that sensations are the foundation of all concepts of the physical world. In *The Science of Mechanics* he proposed a radical reinterpretation of a central idea of Newton. That stone, sitting on the ice, makes a painful resistance when we try to kick it to the far shore. That resistance, that inertia arises, Newton argued, because we accelerate the stone relative to absolute space. No, Mach reasoned, that resistance, that inertia arises because we accelerate the stone relative to the masses of the faraway stars and the other masses of the universe. Einstein groped for a way to relate the two apparently very different identifications of the origins of the stone's resistance: space here, according to Newton; mass there, according to Mach. Two and a half years before he had achieved the final formulation of his theory of gravity, Einstein wrote Mach an appreciative letter. In it he noted that if the forthcoming eclipse observations agreed with his predictions, "then your happy investigations on the foundations of mechanics . . . will receive brilliant confirmation. For it necessarily turns out that inertia originates in a kind of interaction between bodies."

Georg Friedrich Bernhard Riemann

Born September 17, 1826, Breselenz, Hanover. Died July 20, 1866, Selasca, Lake Maggiore.

Space, [in the large] if one ascribes to it a constant curvature, is necessarily finite, provided only that this curvature has a positive value, however small.

It is quite conceivable that the geometry of space in the very small does not satisfy the axioms of [Euclidean] geometry.

The curvature in the three directions can have arbitrary values if only the entire curvature for every sizeable region of space does not differ greatly from zero.

The properties which distinguish space from other conceivable triply-extended magnitudes are only to be deduced from experience.

From "On the Hypotheses That Lie at the Foundations of Geometry," June 10, 1854

Linking Space and Time: Special Relativity

Einstein's geometric account of gravity—also known as the general theory of relativity—could never have come into being had he not first created the special theory of relativity. He made this advance, published in 1905, when he recognized, puzzled over, and finally understood something strange and remarkable about electromagnetic induction, a phenomenon that has no direct connection with gravity.

When the north pole of a bar magnet is thrust through a coil of wire, a momentary voltage is generated between the two ends of the wire. The magnitude of this voltage is predicted by a set of equations formulated by James Clerk Maxwell (1831–1879) and Michael Faraday (1791–1867). On the other hand, when the coil is thrust with equal speed over the bar magnet, the same voltage is predicted—and found—but quite a different equation has to be called on to do the predicting. Any sensible way of looking at this electromagnetic induction, Einstein reasoned, should not make this distinction between which component—the magnet or the coil—was moving and which was stationary. Only relative motion should matter. In his famous paper of 1905 with its innocent title "On the Electrodynamics of Moving Bodies," Einstein postulated that the physics of electromagnetic induction has to look the same whether viewed from the *frame of reference,* or laboratory, in which the coil is stationary or from the frame in which the magnet is stationary.

From the specific case of the coil and magnet, Einstein proceeded to the general postulate: there is no way whatsoever for us to distinguish between two frames of reference in uniform, unaccelerated motion by any difference in the speed of light or in any other law of physics in the two frames. He proceeded to show the revolutionary consequences of this simple feature of our physical world: we have to give up the old idea of space and time as separate aspects of nature. In their place, we have to put spacetime. To put so much into so little, to subsume all of Einstein's teachings on light and motion in the single word spacetime, is to package a wealth of ideas in a small picnic basket that we shall be unpacking all the rest of this book.

The mathematical description of electromagnetic induction by Maxwell and Faraday requires two different sets of equations to calculate the induced voltage, depending on whether the coil is stationary and the magnet moving (left) or vice versa (right). This made no sense to Einstein, as the induced voltage is the same in both cases (assuming all other conditions are identical).

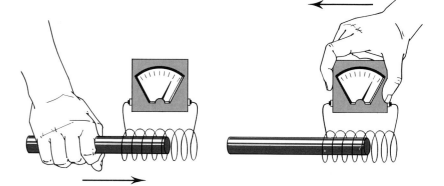

Chapter One

Dozens of other consequences followed from the postulate that physics is the same in all uniformly moving frames of reference. One of the most famous is the conclusion that every form of mass represents a stockpile of energy. Thus when a uranium nucleus splits, the two fragments end up with a total of about one-tenth of one percent less mass than that of the original nucleus. This mass is not lost but rather it is converted into energy. The staggering power of the largest nuclear device exploded to date (60 megatons of TNT equivalent) was released by converting only 2.6 kilograms of matter into energy. Einstein's special relativity pointed the way toward nuclear power, but more important for our understanding of this strange world, it linked space to time in spacetime and pointed to geometry as the key to gravity.

A Link between Gravity and Curvature

Relativity, or the identity of physics in frames of reference in uniform motion, which arose from a question in electrodynamics, proved central to the understanding of electrodynamics. But does frame independence have anything to do with gravity? Let us switch from the old space to the new spacetime and compare the tracks of a moving ball and bullet. As we explore in detail in the next chapter, the curvature of the two tracks, so different when they are pictured in space, turn out to be the same when depicted in spacetime.

This identity of curvature in spacetime was a tantalizing clue. It was a clue to a new view of gravity having to do with curvature. Curvature, yes; but curvature of what? Of the tracks? Of the space? Or of the spacetime, whatever *that* means!

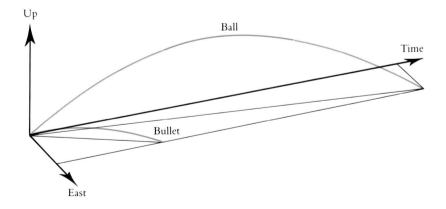

The tracks of a ball tossed across a room and a bullet shot across the same room appear to have very different curvatures; the bullet seems to follow a much straighter path. That is because we naturally visualize the tracks in three space dimensions and ignore the time dimension. But by plotting the tracks in two space dimensions (up-down and east-west) and one time dimension, we see that in fact the degree of curvature of the two tracks is identical. The bullet track will fit exactly over the ball track.

Albert Einstein

Born March 14, 1879, Ulm, Germany. Died April 19, 1955, Princeton, New Jersey.

Einstein's definition of a scientist: "an unscrupulous opportunist."

But the years of anxious searching in the dark, with their intense longing, their alternations of confidence and exhaustion, and the final emergence into the light—only those who have experienced it can understand that.

Albert Einstein

Every boy in the streets of our mathematical Göttingen understands more about four-dimensional geometry than Einstein. Yet, despite that, Einstein did the work and not the mathematicians.

David Hilbert, leader of the Göttingen school of mathematics

The Great Discovery of Free Float

Happily, Einstein happened onto another clue that in conjunction with the curvature clue was to lead him into a new world of ideas. Before continuing the story, let's digress briefly to visit the young Einstein as he began his job at the patent office in Bern in 1902.

Of all conceivable jobs in which to find the man who would lead us to a new world of ideas, was not that of patent office clerk the most unlikely? On the contrary. The head of the office, Friedrich Haller, was a wise man, a kind man, but a strict man who, in Einstein's words, was "more severe than my father . . . [but] taught me to express myself correctly." Listen to Haller's strict instructions to the new clerk: Review each patent application carefully. Check out the working model that comes along with it. Then explain in a single, crisp, clear sentence why it will work or why it won't. And equally clearly, tell me why the application should be granted or why it should be denied.

Day after day for seven years Einstein had to distill out the simple central principle of objects of the greatest variety that man has power to invent. Should we feel sorry for him? No! Rather, let us celebrate the good fortune that so marvellously sharpened his unique talent to unpuzzle the great puzzles! Puzzling about gravity, he could play off in his mind the curvature clue against a new clue given to him by an unlucky painter.

One day in 1908, a painter fell off a roof to the ground. After hearing of the accident, Einstein inquired of the painter how it felt to fall. He learned that the painter felt no weight at all. At that point, there came to Einstein what he was later to describe as "the greatest idea of my life." "In a gravitational field (of small spatial extension) things behave as they do in a space free of gravitation. . . ." Despite this insight, Einstein took another seven years to construct the general theory of relativity. Why the delay? Einstein found it difficult "to free [himself] from the idea that coordinates must have an immediate . . . meaning." Einstein sensed the way ahead. First, hang on to this simple central point: When we fall freely, we cannot detect any effect of gravity in the region near us. Second, chop away "absolute space" and every other mystic concept.

Spacetime Tells Matter How to Move

Two ideas are central to Einstein's conception of gravity. The first is free float. The second is spacetime curvature. What do these terms mean, and how do these ideas fit together?

Einstein's painter was right to believe that he was weightless as he fell. We who stand securely see things the wrong way around because the ground beneath our feet is all the time pushing us away from a natural state of motion. That natural state of motion is free fall, or, better said, free float. To see what free float really means, imagine that the painter falls directly into a shaft that extends straight through the Earth. The painter is never forced away from his natural state of motion by contact with the Earth. Where does the painter get his moving orders for his long, never-ending track through space and time? According to Einstein, the moving orders do not come from some mysterious "gravity" acting in some mysterious way from the center of the Earth or the center of the Sun. Instead, the moving orders come from the geometry of space and time right where the painter is located.

Mass thus gets its moving orders locally. Objects in free float that start off the same—at the same place and time, with the same speed and direction—pursue the same history. They do so because they get their moving orders from the same spacetime. In brief, *spacetime grips mass, telling it how to move.*

Matter Tells Spacetime How to Curve

If spacetime grips matter, telling it how to move, then it is not surprising to discover that matter grips spacetime, telling it how to curve. To understand this corollary notion, let's imagine what free-float spacetime-driven motion would look like if spacetime were not curved. Every object in free float would move in a straight line with uniform velocity forever and ever. The Earth and the other planets would not enjoy the companionship of the Sun. Each would float away on its own proud, disregarding course. Conceivable though such a universe is, it is not the universe that we know. Faced with this difficulty, we could give up the idea that spacetime tells mass how to move. But if we want to retain this idea, despite the observed curvature of planetary orbits and the identical curvature of the tracks of a ball and a bullet through spacetime, we will say, with Einstein, that spacetime itself is curved. Moreover, this curvature is greater at and within the Earth than it is far away from the Earth. In brief, *mass grips spacetime, telling it how to curve.*

Einstein's Geometric Theory of Gravity

The message of Einstein has two parts: spacetime tells mass how to move, and mass tells spacetime how to curve. If these ideas are correct, all physical phenomena must at bottom be local, and physics only looks simple when it is described locally. But, we protest, the Sun undeniably does hold the Earth in orbit; surely that is not local, that is action at a distance. No, every bit of the physics is local, Einstein will reply. The mass in the Sun curves spacetime where the Sun is. This curvature curves spacetime just outside the Sun. That curvature curves spacetime still farther out, and so on. Thus spacetime even as far out as the Earth partakes of a small curvature. Spacetime there, with that small curvature, acts on the Earth, telling it what to do. That spacetime curvature is so slight that the Earth's track is only slightly curved. That is why it takes so long as 365 days for the Earth to complete one revolution in its orbit.

In brief, distant action arises through local law. Each mass follows its natural state of motion, free float, unless deflected by an electric or elastic force. The elastic force of the Earth on our feet drives us earthbound mortals away from the natural condition of free float. The force acting on our feet is not itself gravitational in character. It originates in solid-state physics and the elasticity of matter. We have only to remove that soil, that elasticity, that solid-state physics to be left in a condition of free float.

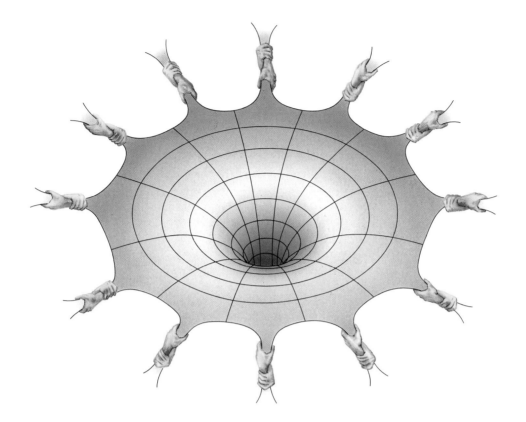

Einstein called his findings of 1905 *special relativity*. He declared clearly, for the first time, that the laws of physics are the same in two free-float frames of reference moving relative to each other. *General relativity* is the name he gave to the later idea that gravitational fall is really free float. Today it is more natural to describe his teaching as Einstein's geometric theory of gravity.

Mass grips spacetime, telling it how to curve, but also spacetime itself grips spacetime, transmitting curvature from near to far.

Is Space the Slave of Mass — or Its Master?

Einstein's declaration that mass tells spacetime how to curve is worth exploring a little further. First, it tells us that spacetime geometry is not a God-given perfection standing high above the battles of matter and energy, but is itself a participant in the world of physics. Second, spacetime being told by matter how to curve, seems to have become the slave of matter. And yet, as Einstein also discovered, the genii of spacetime, summoned forth from the magic lamp to do the bidding of its master, mass, can throw off its chains and bring into action a whole set of dynamic degrees of freedom of its own. Spacetime geometry, in other words, not only can carry "gravitational action" from one mass to another, but can freely on its own transport "gravitational energy" with the speed of light through space. We'll take a closer look at these energy-bearing gravity waves in Chapter 11. Finally, we are beginning to appreciate that mass may be the slave and spacetime, the master. Energy, we have long known, is conserved, and mass, we learn from special relativity, is another way of speaking about energy. Instead of saying that spacetime curvature in and around a region where there is mass behaves as it does because mass-energy is conserved, we are finding it more natural to explain the conservation of mass-energy by the laws of spacetime curvature. We are on the way toward believing that mass would not even know how to behave (see Chapter 7) unless it were "wired up" to space geometry.

Einstein's theory of gravity is thus much more than the story of a master-slave relationship between mass and spacetime. It tells us that spacetime geometry is a new participant on the scene of physics with a dynamics of its own, as real and lively as the dynamics of the electromagnetic field. What Einstein gave us, then, is more than general relativity, more than a geometric theory of gravity; it is a theory of *geometrodynamics*. Different as these three names are, all refer to the same theory, and so I shall use them as synonyms.

Our knowledge of gravity is a road map by which we can navigate the marvels of heaven and Earth, the world of space and time, the valleys and waves in spacetime. It is a guide to traveling over land and sea, through air and space. To prospecting with gravimeter for oil and mineral. To putting the tides to use. To launching rocket, satellite, and spaceship. To deciphering the doings of moon and planet, star and galaxy. To unpuzzling the power of the black hole and the dynamics of the universe.

Newton gave us the first true road map. Leibniz pointed out its weakness—the concept of absolute space. Space geometry, Riemann went on to argue, is not bound by the iron inflexibility of Euclid's axioms but must be curved; acting on mass, it must respond to mass. Mach, on different grounds, reasoned that mass *there* somehow governs inertia *here*. Einstein's 1905 special relativity made clear that any road map to happenings must deal not with space but with spacetime. Einstein first glimpsed how all these ideas could be merged when he suddenly recognized in 1908 that fall is not fall. It is free float.

Newton, Leibniz, Gauss, Riemann, Mach, and many others were all instruments in an orchestra of ideas about gravity and space and time. But Einstein was the director, the one who guided all the individual instruments into a triumphant symphony. The score of this symphony is written in numerous equations. Like the concertgoer who enjoys Beethoven without reading music, we shall explore in this book the beauty and power of Einstein's symphony, without mathematical abracadabra, in straightforward, everyday language, understandable by all. We begin with the opening movement, free float.

We let our boat drift as it would...

and felt that we were

sailing in empty space and

riding on the wind...

we were light as if

we had forsaken the world,

and free of all support

like one who has become an immortal

and soars through space...

Su Tung-p'o (1036–1101)

2 From Fall to Float

Venture far

To see the nearby

With new eyes.

Perceive yesterday's gravity,

Whether acting on man or mass,

As today's free float.

In the movement of the mass

Grasp the message of the medium:

"I, medium that grips you,

Man or mass,

And tells you how to move

Am not space.

I am spacetime."

The weightlessness experienced by astronauts in space was presaged by the Chinese poet Su Tung-p'o (1036–1101) in his description of floating on the Yangtze river.

Let's join Neil A. Armstrong, Edward E. Aldrin, Jr., and Michael Collins in *Apollo 11* as that historic spaceship roars aloft into space. Taking a loop about the moon, we look back from 380,000 kilometers "to see the Earth as it truly is, small and blue and beautiful in that eternal silence where it floats. . . ." Instead of returning to Earth, we leave the Solar System, leave our galaxy itself, and pass through the Virgo cluster of galaxies, 2500 slowly turning groups of stars. Most of these are comparable in size with our own galaxy—the Milky Way—with its hundred billion stars, but they are too far away and too dim for us to see. Thus, apart from a few faint, fuzzy dots, the sky is black. As we press on and on with our voyage, we leave even these dim markers behind. We find blackness ahead, blackness behind, blackness on all sides of us—nothing but blackness. It will be long before chance again throws a few faint pinpricks of light into the all-encompassing night.

Free Float among the Stars and on Earth

From the vantage point of a modern spaceship, the reality of space as the master of motion is easily appreciated.

Space, the empty space that we everyday count as nothing, has to us in *Apollo 11* now become everything. There is no other master but space at hand to tell things how to move. We toss a ball to Aldrin. It goes along a

Chapter Two

line as straight as anyone could want. As it smacks into Aldrin's hand it has as much will as on Earth to keep on going as it was. Where else can it get its moving orders except from space? Only from space can the ball get its orders to go straight, to maintain its speed, to smack as hard as it does any object that gets in its way. Space, theater of motion, is master of motion. Armstrong and Collins join the play, tossing their own ball back and forth, and now and then scoring a hit when their's collides with ours. Again, we all see the grip of space on motion: dictating that each ball moves in a straight line at uniform speed, until its smack against some object results in a change in speed or direction.

Space may be the master of motion within *Apollo 11,* but is it still so outside? To find out, we fire two marbles fore and aft, each at 1 kilometer per second. According to our radar, each marble continues in its original direction with its original velocity as long as the radar echo still shows on the screen. The grip of space on motion is without any limit that we can discover.

It is easy to appreciate the simplicity of motion within a spaceship that is floating among the stars. It is hard to think of conditions being equally simple when we return to Earth. After returning to Earth in *Apollo 11* and waiting to debark, we pass the time by tossing a ball. Does it move in straight and uniform motion as it did a few days earlier in

Any object—a coin, key, or ball—follows a straight track in a free-float spaceship. But when Earth or rocket pushes on the spaceship, the tracks curve relative to the spaceship.

The same ball thrown from the same corner of the room in the same direction with the same speed appears to move quite differently when viewed by an observer in the room when it is not in free float (left) and when it is (right). Yet in the two pictures the ball arrives after the same time at the same location in spacetime. Einstein recognized that fall is illusion. This illusion arises from looking at motion from a reference frame (the room at the left) that is not in free float.

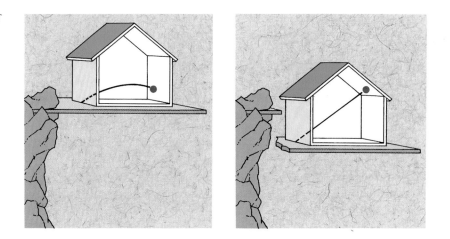

space? No indeed! Like any ball thrown on earth it traces out an arc, an arc so mathematically precise that it appears as if some mysterious force must be acting through space to direct motion. From the Greeks to Galileo to Newton, that motion-directing force—called gravity—was something to be "explained."

Einstein brought a revolutionary new view. He asked what causes a ball tossed across a room on Earth to follow a curved path? Is the cause in the ball? Is it some mysterious force of gravity? Neither, Einstein tells us. It's the fault of us—the viewers—and the fault of the floor that forces us away from the natural state of motion, the state of free fall or, better put, free float. Let the room be cut loose at the moment we throw the ball and the ball will pass through the same space as before. It will go through the same motion as before. It will arrive at the same location in spacetime in the same time. However, the motion will *look* like a straight line, as shown in the diagram above, because we who look at it are now in a different "frame of reference," a *free-float frame of reference.*

The ball's "arc" is an illusion! "Gravity" is an illusion! When we are in free float, neither exists. And the only thing keeping us from the free-float condition is the force of the floor under our feet. As my 15-year-old granddaughter, Frances Towne Ruml, pithily put it:

What's the fault of the force on my feet?
What pushes my feet down on the floor?
Says Newton, the fault's at the Earth's core.
Einstein says, the fault's with the floor;
Remove that and gravity's beat!

As companions of Aldrin, Armstrong, and Collins on *Apollo 11*, we spent weeks in a condition of free float. Living among the stars helped us to appreciate the grip of space on every moving object. That grip keeps mass moving in a straight line. That grip keeps speed constant. That grip makes a moving mass deliver a smack to anything that would alter its velocity. If space is so all-powerful in the blackness that lies between the galaxies, why should its grip be any less powerful on Earth? Why should the rules that space enforces here on Earth be any less simple than out there among the galaxies? This all seems quite reasonable. So let's imagine free float here on Earth, inviting along a friend, Tim, who was not with us on *Apollo 11*.

A few seconds after we enter our "experimental" free-float room, Tim grabs the doorjamb, terrified. "We're falling!" he cries out. His fear turns to astonishment when we tell him not to worry.

"A shaft has been sunk through the Earth," we tell him. "It's not the fall that hurts anyone. What hurts is whatever stops the fall. But nothing stands in our way. All obstacles have been removed. Free fall," we assure Tim, "is the safest condition there is. That's why we call it free float."

"You may call it float," Tim says, "but I still call it 'fall'!"

"Right now that way of speaking may seem reasonable," we reply, "but after we pass Earth's center and start approaching the opposite surface of the planet, the word 'fall' may seem rather out of place. Then, I hope you'll find 'float' is a better word."

Several of us start to toss a ball. The ball flies across the free-floating room on a straight course, just as the ball did on *Apollo 11*. Motion in a straight line and at uniform velocity for marbles, pennies, keys, and balls is the norm in our free-float room. This simplicity is rigidly maintained by the grip of space on motion. There are no jolts, no shudders, no shakes at any point in all the long journey through the shaft from one side of the Earth to the other.

Free-Float Reference Frames

What first strikes us about the concept of free float is its paradoxical character. As a first step to explaining gravity, Einstein got rid of gravity! No more evidence of gravity is to be seen in the freely falling room than in an unpowered spaceship or any other free-float frame of reference.

The second feature of a free-float frame of reference is its local character. Let's arrange to cut loose, at the same time, three free-float rooms, one from a cliff in Kalamazoo, Michigan, one at nearby Kokomo, Michigan, and one from a cliff in Kyoto, Japan. The first two fall in only slightly different directions, but the third falls in a totally different direction. Free fall *here* and free fall *there* are not the same. The concept of a free-float frame of reference does not apply outside a limited and local region. Einstein recognized earlier than anyone else that the first step toward a simple description of motion is a local description of motion. As Stephen Shutz put the contrast between Isaac Newton's view of gravity and Einstein's, "Rather than have one global frame with gravitational forces, we have many local frames without gravitational forces."

A free-float frame of reference has a third feature. Being responsive to space right where it is, it bears witness to the grip of space on everything that goes on within it. Nowhere else but from space itself do objects—pennies, balls, spaceships, cannon balls, crashing cars—get their number-one moving orders. Nowhere more simply than in a free-float frame of reference can we see the story of motion, of space and time, and of that strange unity between space and time that we call spacetime.

At this point we seem to be far removed from the notion that gravity is a manifestation of the curvature of spacetime. No gravity! No curvature! Every free object moves in a straight line at uniform velocity. Simplicity we have achieved, yes, but isn't it too much simplicity? Haven't we simplified out of existence the very gravity we wish to understand? Not at all! There is no way to get into Einstein's geometric picture of gravity—the notion of spacetime curvature—except by the simplest way there is, by looking at local conditions first.

What does the student of shape and form do when she wants to get an exact account of the shape of a Ford fender, and how it differs from a Fiat fender? Take endless measurements with assorted meter sticks running this way and that? No, she tells us, the right way and the simple

way is quite different. Paste a postage stamp down on the metal surface. One postage stamp in and by itself gives no evidence that the fender is not flat. Paste down a second postage stamp. Again, no hint that the fender is anything but flat. Only after a good part of the fender's surface is covered with rectangular postage stamps do we see that they simply cannot be laid down in a regular grid covering the entire surface as they could be on a flat surface. The departure of adjacent postage stamps from regularity provides a local, measurable, and simple description of the fender's curvature. The greater the curvature, the greater the departure.

So too with "gravity as manifestation of the curvature of spacetime." We cannot see any evidence of spacetime curvature or of gravity by pasting down on spacetime that single "postage stamp" which we call a "local free-float frame of reference." Only in the discrepancies between nearby "postage stamps," in the discrepancy between the nearby Kalamazoo and Kokomo free-float reference frames, will we be able to see clear evidence of that curvature (very different from the more immediately apparent curvature of the fender's surface) that Einstein had the wit and wisdom to recognize as the essence of gravity. If it is easier to begin with a postage stamp than with the entire automobile fender, it

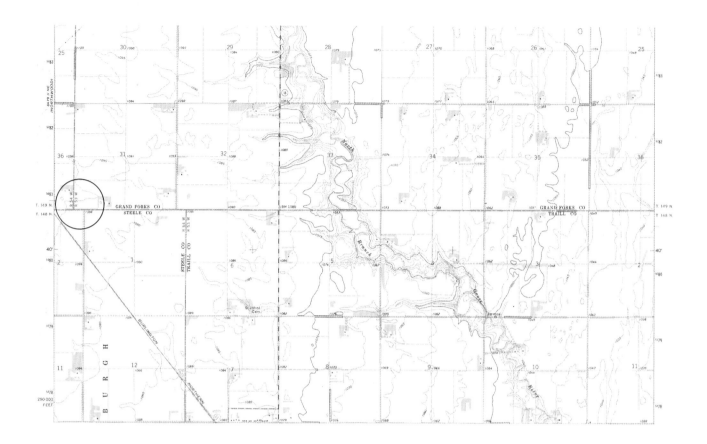

Local evidence of Earth's curvature: neighboring 10-mile-square townships in North Dakota will not line up.

will also be easier to look at motion and a local free-float frame of reference in the rest of this chapter and the next two before we turn to "gravity as spacetime curvature" and all its consequences.

Free Float: The Same for Materials of All Kinds

Galileo Galilei (1564–1642), according to a popular but unconfirmed legend, dropped weights of different sizes and materials from the Leaning Tower of Pisa and found that they hit the ground at the same instant. About three centuries later, Baron Roland von Eötvös (pronounced 'ut vûsh) (1848–1919), after some thirty years of experiments in Budapest, succeeded in establishing Galileo's conclusion with the then unprecedented accuracy of five parts in a billion. He and his colleagues D. Pekár and E. Fekete showed in effect—by an indirect but ingenious arrangement—that materials as different as wood and metal, allowed to float freely, will pursue identical tracks through space and time. If their tracks were different, then there would be one free-float frame of reference for wooden objects and a different free-float frame of reference for metal objects. It would be impossible to claim that objects get their moving orders from spacetime itself and nothing else. Einstein's concept of spacetime as the master of matter would collapse. Therefore the Eötvös experiments and their even more precise modern-day successors are of central importance.

The year 1987 brought a brief flurry of favor for a "fifth force." Near the Earth, experiments of one group seemed indirectly to show that materials of different kinds do not fall in step. Other groups thereupon carried out other experiments and got discrepant results. The more precise such experiments have become, the smaller the purported differences—if any—in the free fall of different materials have turned out to be. At this writing honesty compels me to say that there exists no battle-tested evidence whatsoever for the existence of any such fifth force, any departure from identical free fall for all things.

One of the most distinguished philosophers of science of our century, Sir Karl Popper (1902–1989), argues that experiments can never prove a theory; they can at most disprove it. Even to begin to win belief, a new theory has to account for all the facts, and have all the predictive power, that the old theory did. How much belief it can win depends on the scope of its new predictions, on how many unexpected ways it can expose itself to destruction—and on never failing a single test.

Galileo Galilei

Born February 15, 1564, Pisa, Italy. Died January 8, 1642, Florence, Italy.

If one lets fall simultaneously from the same height two bodies differing greatly in weight, one will find that the ratio of their times of motion does not correspond to the ratio of their weights, but that the difference in time is a very small one.

John Philoponus of Alexandra, in his Physica, *517 A.D.*

Refering to the painting by Sustermans shown here, now in the Uffizi Gallery: My portrait is now finished, a very good likeness, by an excellent hand.

Galileo Galilei, September 22, 1635

When shall I cease from wondering?

Galileo

EÖTVÖS'S EXPERIMENT

To measure the travel time of a falling body more precisely demands more time. So realized Baron von Eötvös, who therefore devised an experiment of a new kind that offered unlimited time for measurement. It focused on the central issue: Is there for any substance any such distinction, as old writers assumed, between its "gravitational mass," on which the center of the Earth is conceived to pull, and its "inertial mass"? The "inertial mass"—today simply "mass"—resists being set in motion or, if already endowed with a velocity, resists any change in the magnitude or the direction of that velocity.

Imagine an object endowed with inertial mass but no gravitational mass, hanging at the end of a fiber and carried around and around in a circle by the spin of the Earth. With no gravity to hold it down, it will pull the fiber straight out from the axis (red line in the diagram). On the other hand, an object that has no inertial mass at all will not be thrown outward from the axis of spin of the Earth. Instead—if it has any gravitational mass—it will tug the supporting fiber to a position where it points straight down to the center of the Earth (blue line). An object deprived neither of gravitational mass nor of inertial mass will pull the supporting fiber neither straight out nor straight down, but instead to an angle of uprise (green line). Does the ratio of inertial to gravitational mass differ from one substance to another? Then the angle of uprise will differ, too. And for measuring it there's all the time in the world!

This beautiful idea Eötvös put into action. He and his colleagues, in experiments extending over some thirty years, on a variety of substances, were able to establish that there is an angle of uprise: that angle is greatest at latitudes 45°N and 45°S, and amounts there to a tenth of a degree. They found to an accuracy of 5 parts in 10^9 that the angle of uprise was identical for every substance tested.

Baron Roland von Eötvös

Born July 27, 1848, Budapest. Died April 8, 1919, Budapest.

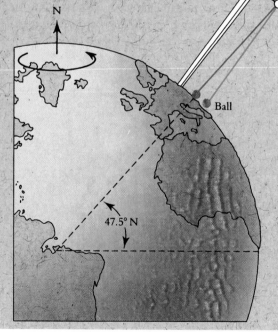

N

Ball

47.5° N

The experiment of Eötvös and his Budapest colleagues.

Isaac Newton's theory of motion could have survived if a ball of wood and a ball of lead had been found to fall at different rates from the Leaning Tower of Pisa. Einstein's theory could not. According to Newtonian theory, the acceleration of a falling object is governed by the ratio of two effects: (1) the "gravitational pull of the Earth" on the "gravitational mass" of the object and (2) the resistance set up by the "inertial mass" of the object. If all objects "accelerate downward" identically, then in Newtonian theory, a Newtonian would say that the ratio of "gravitational mass" to "inertial mass" is identical for all objects. If they fail to "accelerate downward" identically, then, with a shrug of the shoulders, he would say that the ratio of "gravitational mass" to "inertial mass" happens not to be the same for all objects—interesting observation, but no threat to the Newtonian theory of motion.

Not so in Einstein's theory! In it there is no "gravitational pull" from a distance. Matter gets its moving orders from space right where it is. There is no room for any difference in "rate of fall" between a wooden ball and a lead ball. By making that prediction Einstein's theory exposed itself to destruction as Newton's theory never did. Moreover, it survived that test—and tests of its predictions on half a dozen previously unexpected and non-Newtonian effects.

Einstein's theory is regarded today as the viable theory of motion. Newton's is not. With the demise of Newton's theory of motion, the terms "inertial mass" and "gravitational mass" are slowly fading into antiquity. Does Einstein's theory use the concept of "mass"? Yes, but *not* "inertial mass" and "gravitational mass."

The Medium That Sustains Free Float: Space? Or Spacetime?

The guilt for gravity? Where pin that guilt? On the "falling" object? Or on the medium in which it does the falling? If every object "falls" the same, if the purported distinction between "inertial" and "gravitational" mass collapses, then the finger of guilt for "fall" points, not to mass, but to medium. Once Einstein grasped the concept of free float in 1908, he had taken a substantial step toward the final idea: Gravity is not a foreign and physical force transmitted through the medium that surrounds us. It is a manifestation of the curvature of that medium.

What is this medium whose curvature is responsible for gravity in some way that was mysterious in 1908? Space? Or spacetime? To find the answer, we do not have to look into some learned treatise but only to look with new eyes at the throw of a ball.

So we begin launching balls from a mechanical device on one side of the room to a constant target point on the other side. To help get a new

Robert H. Dicke

Born May 6, 1916, St. Louis Missouri. Dicke is an American physicist and pioneer in microwave physics and superradiance, the precursor of the laser. In 1964, Dicke and his younger colleagues, Peter G. Roll and Robert Krotkov, by a delicate, ingenious, and weeks-long experiment, established to the unprecedented precision of 1 in 10^{11} parts that aluminum and gold—two substances of very different density, atomic number, and nuclear composition—fall at identical rates. Once again, Einstein's notion that the medium, not the mass, determines motion survived experimental test.

view of things, we bring along two photographers, Charlie and Madeline, with movie cameras. Charlie has a wide-angle camera and stations himself along the middle of a wall running parallel to the ball's path. Madeline shoots her pictures from a point out on the porch, behind a shock-resistant plexiglas window, aiming head on toward the oncoming ball.

We watch ball after ball zoom across the room from catapult to target point. Fast ball—short time of passage and only a slight rise of its trajectory at midpoint. Slow ball—three times as long to cross the room and nine times the rise at midcourse. The two tracks are totally different in curvature when pictured against the backdrop of the wall, chalked onto that vertical plane, pictured in space. When we look at the tape from Charlie's sideline camera, the pictures confirm this naked-eye view.

In left-right, up-down two-dimensional space, we see that different balls give different curves. Conclusion? Space pictures of these curvatures are bad. Why? They point the finger of suspicion for gravity the wrong way, toward the ball. Different balls, different curves. Ergo, the cause of gravity must lie in the ball. But that is wrong. The equivalence of free fall to free float—whatever the substance of the "falling" object—tells us once again that the cause cannot be located in the ball. It must lie in the medium. So once again we ask ourselves *what* medium? If not space, then it must be spacetime. But doesn't spacetime have four dimensions? How in the world are we ever going to be able to picture that?

Then we remember that Madeline was filming from a different view—from the porch looking head on at the balls moving toward her. Her films reveal the position of the balls in only one space dimension—

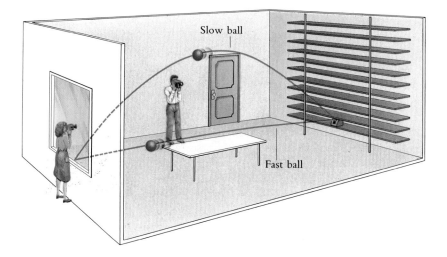

When viewed from the side with the naked eye or a movie camera, the tracks of fast and slow balls are very different.

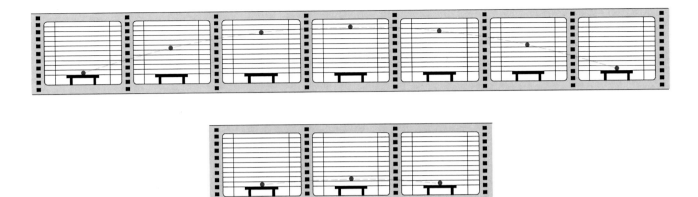

height, but each successive frame shows height at a different instant in time. There it is, a record of motion in up-down, past-future two-dimensional spacetime! In this two-dimensional spacetime, the tracks of the fast ball and the slow ball can be superimposed: they have the same curvature. Spacetime is the medium responsible for the motion of mass!

The Addition of Time to Space

Our experiments comparing the paths of balls enable us to visualize the medium of spacetime—at least a two-dimensional flat spacetime. But how can we grasp the concept of spacetime curvature? Earlier we saw that the curvature of an automobile fender was revealed by the inability to lay down rectangular postage stamps in register on its surface. We can gauge the "imperfection"—the curvature—of a two-dimensional space by seeing how perfect pieces, postage stamps or ideal townships, fail to fit. Perhaps the same approach will work with spacetime. In 1908, Einstein recognized that local free-float reference frames are the spacetime equivalent of ideally rectangular postage stamps or ideally square townships; they are, in other words, "perfect pieces of spacetime."

We are now almost ready to follow Einstein's 1908–1915 advance into spacetime curvature. About to start, we suddenly realize that our images of a free-float frame do not yet measure up to what we need. Room cut loose from a cliff—good. Ball coursing through that room on a course straight in space—good. But where is *time,* and that unity of time and space that is spacetime? The track of the ball is straight not only in space, but also in spacetime. This straightness signals the flatness of both space and spacetime, at least so far as room-scale evidence reaches.

Looking head on as the balls approach, the movie camera records their height at each successive instant in time. To see the curvature of the tracks in spacetime, we cut out each frame and line them up side by side in sequence, placing the fast-ball frames below the slow-ball frames. The tracks of fast and slow balls in this two-dimensional up-down, past-future spacetime have the same curvature, as appears on comparison of the identical tracks of the fast ball and the slow ball in the three middle frames. The tracks of the balls curve differently in space (illustration on page 28), but in spacetime (shown here) the tracks curve the same.

Flat spacetime was the essence of Einstein's special theory of relativity of 1905. He had that idea firmly in hand when he started in 1908 to move forward into the domain of curved spacetime. This unity of space and time was soon being expounded by Einstein's former Zürich mathematics professor, Hermann Minkowski: "Henceforth space by itself, and time by itself, are doomed to fade away into mere shadows, and only a kind of union of the two will preserve an independent reality." Before we proceed to the curvature of spacetime and gravity, let's see how nature packages space and time into the unity of spacetime.

How, then, do we "add time" to our free-float room cut loose from the cliff? First, we can install fixed clocks all over the free-float room, to give a time that our friends in the business generally call *laboratory time* (lab time) or *frame time*. Second, we clip to the flying ball a beacon flashing at standard intervals. The count of the flashes from the beacon registers *proper time,* the "wristwatch reading" of the flying ball. Install, too, identical beacons, identical wristwatches, identical measures of proper time, on all the other objects that we bring into the room, each flying freely this way or that on its own straight-line course.

Note that we cannot—as we did in the earlier experiment—add time by installing a camera to observe all this. Why no camera? Isn't a camera an impartial observer? And doesn't the word *observer* occupy a central place in every discussion that anyone has ever heard of Einstein's special relativity, his spacetime? That innocent but ambiguous word has created endless confusion for more than half a century. It fails to distinguish between two totally different concepts of observer. One is right. The other is wrong. The camera epitomizes the wrong idea—wrong because it is not located where the action is.

The right idea of observer is more sophisticated. The observer is not somebody who functions at one location. He is someone who—with the help of an army of registering clocks—operates all over the place. He is the person who collects the printout from all these recorders.

The observer, taken in this sense, capitalizes on the concept of lab time. He makes use of a great latticework of "clocks," all fixed in a free-float frame, all ticking away in synchronism. Each is equipped with a printer. Do two flying objects collide right next to the site of one of these clocks? That clock's antennae pick up this event, and the clock prints out a card that reports the event. The card tells in what column— say the column of clocks 24 meters to the right of the left-hand wall—the clock stands that registered this event. It tells in what row—say the row of clocks 7 meters above the floor—the clock is located. Finally, the time of the event is recorded on the card.

Now we have our brand-new free-float room equipped with its latticework of fixed clocks. We toss a few balls leisurely back and forth.

A CAMERA CAN SO LIE

The camera is only a little wrong, even negligibly wrong, when its successive frames register the position of an object moving at low or negligible velocity. In this case, the correction for the time required for light to fly from the object observed to the camera film is negligible. The camera is totally wrong, or a dismaying source of complication, however, when it registers light signals from an object moving with a speed comparable to the speed of light itself.

How to find such an object? Radio telescopes provide successive pictures of a super-fast jet of matter emerging from a distant galaxy. The lead point of the jet comes ever closer to us night by night. Remarkably, when astronomers first measured the distance such a jet had traveled in one day, then calculated the speed by dividing the distance by one day's time, they found that the jet was approaching us at a speed faster than light! Could Einstein have been wrong? The first interpreters did not make any mistake in measuring how far the jet moves from night to night. But they were off in thinking that one day of our time was one day of jet time. They had not allowed for the ever-shortening time it takes for light to come from the jet tip to us. While traveling that measured distance, the jet's proper time was greater than the time recorded by the astronomers. We had received the information on the jet's later position sooner because the messenger, light, had less distance to travel. The telescope's camera was wrong because it was not located where the action is!

The 300,000 light-year long jet of matter ejected from the active core (upper left) of the galaxy known as NGC 6251 and recorded by radio telescope (color in printout indicates intensity of radio-wave emission). This particularly impressive jet has an apparent velocity of three-tenths of the speed of light, but at least 23 such jets show apparent velocities greater than the speed of light, some of them 10 times as great.

We find that the clocks are registering only a few events. Life in that room seems tame. Closer scrutiny, however, discloses a fantastic world of activity: atoms zipping to and fro; bullets of light, quanta of radiation, photons being given off here, absorbed there; and collisions between this and that all over the place. No observer could ever hope to record more than the tiniest fraction of these events. That task, however, a truly alert army of printout clocks—as we conceive it in imagination—accomplishes for him. The observer simply picks up the printout cards.

What is he going to do with these cards? Give them to us. What are we going to do with them? Study them to understand what has been happening on this battlefield of motion. Well then, we say to him, why don't you go home? Leave the printout cards, the battle scars of mass in motion, right where they accumulate. As we stroll around the battlefield afterwards, we can just as well view them one by one ourselves.

Conceiving matters in this light, we say goodbye to the card collector. As he goes, he smiles, peels the label Observer off his cap, and hands it to us. We tack it up over the army of printout clocks. No agency more than it deserves the title Observer. No agency is more far-flung. None can outdo it. This ideal observer, this Einstein observer, this special-relativity-frame-of-reference observer, lets us know anything that goes on anywhere.

Anywhere? Anywhere in the Kalamazoo free-fall room. Or anywhere in the Kyoto free-fall room. But not anywhere in both. Not anywhere in a domain so extended that perceptible differences develop between one object and another in their free fall. In brief, the observer operates in an inertial reference system that is local.

Local the free-float frame of reference is, but it is not unique. We ask a friend, Marie, to set up her own great army of clocks, her own latticework of printout devices, her own metric array of registers. Marie anchors what she calls the "lower" left-hand corner of her latticework at what we call the "lower" left-hand corner of ours. That one precaution is not enough to guarantee identity of the two recording systems. Marie's, we discover, is tilted relative to our own. The consequence of this tilt is illustrated in the diagram on this page. In our observing system, the

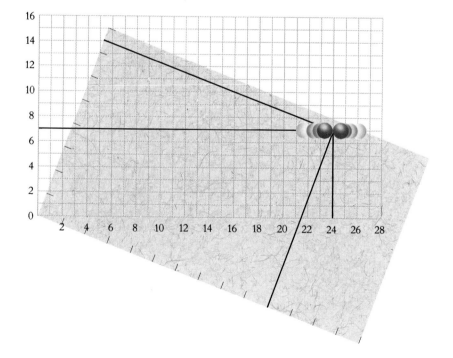

The collision of two balls is registered at different coordinates in two different reference frames that are tilted relative to each other.

collision of two balls registers as 24 meters to the right, 7 meters up. The same collision, in Marie's system, prints out as 20 meters to the right and 15 meters up.

Great differences can thus exist between the "coordinates" of the same event as registered in two distinct reference frames. Coordinates so much affected by someone's whim in orienting his lattice of clocks? Then those coordinates cannot have any truly fundamental bearing on the description of motion. How then are we to describe the "where" of events that will be independent of frame orientation?

Trying one idea after another, we discover a striking coincidence. Take Marie's right reading for the collision. Square it. Take her up reading for the place of collision. Square that number, too. Add the two squares. Miraculously, the result comes out to be the same figure that we get when we apply the same mathematics to our own very different readings.

Marie's reference frame:
$$(20 \text{ m})^2 + (15 \text{ m})^2 = 400 \text{ m}^2 + 225 \text{ m}^2 = 625 \text{ m}^2 = (25 \text{ m})^2$$

Our reference frame:
$$(24 \text{ m})^2 + (7 \text{ m})^2 = 576 \text{ m}^2 + 49 \text{ m}^2 = 625 \text{ m}^2 = (25 \text{ m})^2$$

In mathematical language we have discovered a quantity that is independent of, or—more powerfully stated—*invariant* with respect to, the orientation of the coordinate system. We call this quantity *distance*. Thus distance is invariant with respect to orientation of the reference frame.

Felix Klein (1849–1925) enunciated in his 1872 lectures at the University of Erlangen what was to prove one of the most fruitful theses in all of mathematics: one kind of geometry is distinguished from another by the kinds of quantities that are invariant in it. We have just come upon the quantity, the invariant, that characterizes Euclidean geometry. The Babylonians, the Egyptians, and the Greeks recognized that distance is invariant regardless of the orientation of the reference system.

But what is invariant in spacetime? *Interval.* Einstein realized in 1905 that interval is analogous to distance. The difference? Distance measures separation of two points in space. *Interval measures separation of two events in spacetime.*

We've just seen that comparison of Marie's readings and our readings for the position of the colliding balls reveals the invariant quantity distance. To reveal the analogous invariant quantity—interval—we only have to be more daring in exploring the idea of free float. And more daring we shall be in the next chapter. On to interval!

3

Interval: Revelation That All of Space Is Ours

Oh event,
Sparkling grain of sand
On the fabric of existence,
Oh interval,
Gossamer tie
Between event and event,
You two tear away the clouds
Of "absolute space" and "absolute time"
And reveal to us spacetime—
Spacetime as doorway,
Doorway, daring traveller,
To the enormity
Of space and time
Open to our visitation.

Rays of light tie near and far. Light makes zero-interval links between event and event throughout the universe.

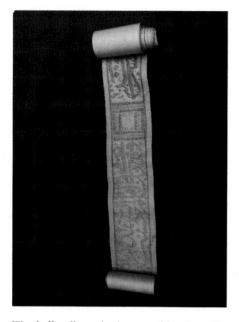

We shall roll up the heavens like the rolling up of the scrolls by a scribe. As We began the first creation, so shall We repeat it.

Koran, *Chapter 21, al-anbiyā' (103)*

Think of spacetime as a great record of all that was, is, and evermore shall be. Spacetime, like a great unrolling papyrus scroll with densely sprinkled grains of sand glued to it, loaded throughout its vastness with microscopic events, the collision of particle with particle—or of particle with that bullet of light we call a photon. These events are connected by the straight-line tracks of particles and photons as one grain of sand on the scroll is connected to another by the faint strand of a spider's web. We don't need coordinates on the papyrus to be able to pick out and point our finger at an extra-bright ruby-red grain of sand. We don't have to know how to measure space and time to point to the impact of a meteor on Arizona as an event. Space and time are loaded with events as surely as the scroll is sprinkled with grains of sand. These events impart to spacetime a definiteness that rises above all coordinates, all frames of reference, all clocks, all measures of time. The right order of ideas is not first frames of reference, then spacetime. The right order is first spacetime, then frames of reference.

How do we measure the spacing between points, between events, between grains of sand? We know from schooldays how to define and measure the separation—or distance—between two grains of sand:

$$(\text{distance})^2 = (\text{east-west separation})^2 + (\text{north-south separation})^2 \\ + (\text{up-down separation})^2$$

Between event and nearby event, there is a separation as definite as that between one grain of sand and another on the papyrus, but of a different quality. Events in the real world are separated in time as well as in space. A separation of this new kind cannot be called distance. Separation in spacetime is called *interval*. The difference in quality between grains of sand and events in spacetime shows most dramatically in this, that the interval for some pairs of events is spacelike, for other pairs timelike, but never both.

How can we tell whether an interval is spacelike or timelike? We look at the east-west separation between the events, the north-south separation, the up-down separation, but also the separation in time. We square all four quantities, then subtract the time squared from the sum of the other three squares. When the difference is positive, it gives directly the square of the interval between the two points and tells us more—that the interval is spacelike:

$$(\text{spacelike interval})^2 = (\text{east-west separation})^2 + (\text{north-south separation})^2 \\ + (\text{up-down separation})^2 - (\text{separation in time})^2$$

In contrast, the difference, when it comes out negative, signals that the interval between two events is timelike:

$$\text{(timelike interval)}^2 = \text{(separation in time)}^2 - \text{(east-west separation)}^2 - \text{(north-south separation)}^2 - \text{(up-down separation)}^2$$

The doorway to the discovery of interval was innocent: uncovering the place of light in the scheme of things.

From Speed of Light to Invariance of Interval

That the speed of light is the same in every free-float frame of reference; that space and time are inseparably united in spacetime; that interval is the measure of separation in spacetime—this triplet of revolutionary ideas, this 1905 special-relativity package deal, Einstein was forced to accept as the only possible way to make sense out of the puzzles then floating in the air.

Speed of light the same in every free-float reference frame? Preposterous! Perhaps not. Let's rejoin *Apollo 11,* as it floats among the galaxies, to examine this strange idea. A free-float rocket zooms past us to the right at 80 percent the speed of light. From this rocket, a flash of light is launched, also to the right, at the full speed of light. Relative to us, does not that flash fly to the right at 1.8 times the speed of light? No! It has the same speed relative to us in *Apollo 11* as it does relative to the rocket.

Einstein's doorway to this fantastic discovery, as we saw in Chapter 1, was electromagnetic induction. Since the same voltage is generated when a magnet is moved relative to a stationary coil and when the coil is moved relative to the stationary magnet, Einstein concluded that only the relative motion of coil and magnet matters. Einstein's analysis forced him to a further conclusion: in all inertial frames, in all free-float frames of reference, light has the same speed and the laws of physics have the same form. In brief, nature offers us no means whatsoever to distinguish one inertial frame of reference from another, either by any measurement of the speed of light or by any other local "closed-box" physics experiment.

How sure are we that the laws of physics and the speed of light are the same in every free-float frame? How sure, in brief, that Einstein's special relativity is right? Hundreds of experiments and theoretical studies conducted since Einstein's Bern days have not uncovered a single sustainable point of logic or a single battle-tested bit of observational

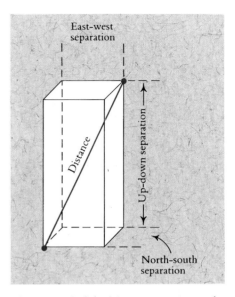

Two events (red dots) in space are separated by a distance that can be calculated from the separations in the east-west, north-south, and up-down directions.

evidence that puts relativity in question. To the contrary, hundreds of interesting consequences have been found to support it. Beyond all the practical consequences of relativity, beyond all the power it gives us to state the laws of physics in simple form, beyond spacetime—the medium of all matter and all change—lies what? The absolutely central concept of the invariance of interval.

Of intervals are the threads made that tie together spacetime. By intervals, directly or indirectly, is every law of physics defined. The interval between one event and another in spacetime has an invariance that defies all attempts to change it by any change in the free-float frame used to measure it.

How best to illustrate interval? Atoms? Too small for everyday experience. Space journey? Too much of a never-never land flavor. As I struggle for an illustration of interval that combines simplicity with vividness, the phone rings. My friend Kip, a fellow physics professor and Einstein enthusiast, greets me and invites me to join him for the Rose Bowl parade the following day. Stumped in my efforts to explain interval, I accept his invitation, and we agree on a place and time to meet. The next morning, I'm off early, putting relativity and interval behind me for the day—or so I thought.

Rose Bowl Parade at Normal Speed

As Kip and I stand along the Rose Bowl parade route, float after gorgeous float passes by. Suddenly one depicting Punch and Judy playing croquet catches my attention. By eye I guess it to be 3 meters wide and about half as long. Punch stands on the sunny side, on the verge of falling off, whacking a great croquet ball toward Judy. She whacks it straight back, standing though she does perilously close to the shady edge, in danger of falling into the street below. Back and forth the ball goes, never stopping for an instant.

"Look Kip," I say, pointing to the Punch-and-Judy float, "what a wonderful illustration of two reference frames in action. The croquet ball makes no forward progress at all relative to the float. But relative to the ground, it is moving ahead at a good clip. Between one whack that Judy gives it and the next, how far forward do you think it progresses?"

"Let's find out," Kip replies. "I'll station myself across from where Judy gives the ball a whack and you run ahead from there to the place where she delivers the next blow."

"Good idea," I agree, scurrying along in step with the float as it glides forward. After I mark the place where Judy strikes the second

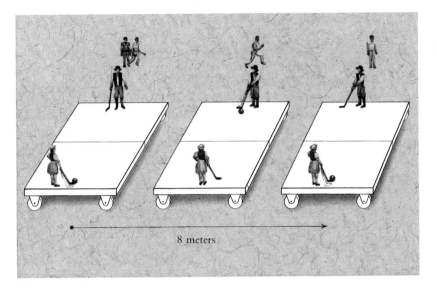

8 meters

blow, I manage to pace off the distance from my position to Kip's despite the press of spectators all around me. "Eight meters," I report.

We repair to a sidewalk cafe for cappuccinos and palaver. "How lovely," I smile at him, "to have such a beautiful example. It's quite clear that the ball, relative to the ground, goes faster than it does relative to the float."

"Furthermore," Kip adds, "we can even say how much faster. Between the two whacks that Judy gave, the ball—and float—had moved forward 8 meters. For the ball that's one round trip. On its one-way trip, then, it goes forward only half as much, that is, 4 meters."

"But that one-way trip, relative to the ground, runs slantwise," I remind him. "It's the hypotenuse of a 3:4:5 right-angled triangle."

"So," Kip continues, "relative to the ground the croquet ball goes 5 meters, whereas relative to the float, it only goes 3 meters. That means it has 67 percent greater speed on the ground. And it would impart a much greater sting to our palms if we—rather than Punch or Judy—ever tried to catch that ball."

Many a thought goes through our minds. The difference between what's going on in the Punch-and-Judy frame and in our ground frame all looks so straightforward, doesn't it? But replace the back-and-forth ball with a back-and-forth flash of light, and we will surely shock our friends with the surprise of relativity. After all, how can anyone learn anything new who does not find it a shock?

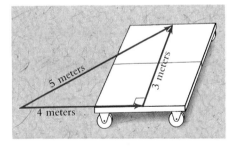

Since the distance between Punch and Judy is about 3 meters, during a one-way trip, the ball moves crosswise 3 meters while the float moves forward 4 meters. Thus relative to the ground, the ball moves 5 meters during a one-way trip.

Rose Bowl Parade at the Speed of Light

Flash of light. Let it be born in the spark at Judy's feet. Let it fly out to a mirror at Punch's feet. Then back again it flies to a photodetector at Judy's feet. Straight crosswise Punch and Judy rate and rank the passage of the flash. *Almost* straight crosswise we also view it. Why? Because the forward speed of the float is so small compared with the speed of light. Kip and I decide to speed up the float in our imagination, speed it up so much that the lightning-fast round trip, Judy to Punch to Judy, ends up in our ground-based frame of reference 8 meters forward of where it started. Then in the course of one round trip from spark to click, the flash will travel two slantwise legs of 5 meters each, a total of 10 meters. Ten meters of light-travel time. Judy insists that the out-and-back trip took 6 meters of light-travel time. We say just as insistently that it took 10 meters of light-travel time. Six; ten; six; ten . . . !

"Our friends probably will insist that time is time is time," I say. "That our clock and Judy's clock are identical and therefore must register identical time between spark and photodetector click."

"Then we will just have to show them the consequence of that line of argument," Kip says. "The light, like the croquet ball, travels in the frame of Punch and Judy 6 meters; in our frame, 10 meters. To argue the same time for different distances is to argue different speeds of light, different laws of physics in two frames; that is contrary to everything we know!"

Carolee, a physics graduate student who had joined us as we entered the cafe, shows signs of headache. "Let me review what you two have been saying." On the back of a placemat she sketches out the top table on the facing page.

"You two," Carolee says, "with all your talk of numbers! Why can't you focus on harmony instead of disharmony. You've been telling

When the float moves forward at 80 percent of the speed of light, then the time between flash of spark and click of photo detector—6 meters of light-travel time according to Judy on the float—will register to ground-based spectators as 10 meters of light-travel time.

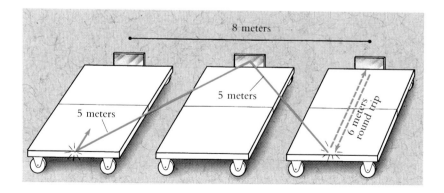

Light travel in two frames of reference

	Punch-and-Judy float greatly speeded up	Our ground frame
Round-trip travel of flash of light	6 meters	10 meters
Time it takes for flash to travel from spark to click, both at Judy's feet	6 meters of light-travel time	10 meters of light-travel time
Distance between positions at which spark and click occur	0 meters	8 meters

us in class that interval is figured like distance. O.K., let's square the numbers, as you instruct. Only don't add the squares, you warn me; subtract them. O.K., let me show you how nicely the numbers turn out." Carolee works out the calculations in the table below.

"Now you can see that the two reference frames don't agree on the separation in space between spark and click. They don't agree on the separation in time between spark and click. But they do agree on the interval between spark and click. Then why don't you two stop chattering so much about those space and time readings that depend so much on our choice of reference frame? Why don't you talk about what really counts, the interval? There's a beautiful feature of the interval between event and event as I now grasp it: it is totally independent of reference frame. Hurray for interval!"

Calculating the interval in two frames of reference

	Punch-and-Judy float greatly speeded up	Our ground frame
(Separation in time between spark and click)^2	36 square meters	100 square meters
(Separation in space between spark and click)^2	0 square meters	64 square meters
Difference = (timelike interval)^2	36 square meters	36 square meters
Timelike interval itself	6 meters	6 meters

Light as Clock

Hans, a friend of Carolee's, had joined our little group as Carolee was writing out her first table. A history student, he was looking bewildered by this time. "All of you physicists are crazy. You make such a to-do about the speed of light, always saying that nothing can go faster than light. Yet right here in this first table you show me that the ground frame moves forward 8 meters during the time that light travels only 6 meters. Have I gone mad? Or you?"

"You are sharp, Hans," I reply, "but this time you have let your passion for detail hide from you the central point. You took our *ground*-based figure for forward motion, 8 meters, and put it together with the *float*-based figure for time, 6 meters of light-travel time. You divided the one by the other to get speed. Why not take what I paid for fuel for *my* car last week and divide it by the number of gallons you bought today for *your* car, to figure the cost of a gallon of fuel? A crazy, mixed-up, wrong way to work out cost—but no crazier than your way to figure speed!"

"If I did it wrong," Hans manfully replies, "tell me how to do it right. What is the right figure for the velocity of the speeded-up Punch-and-Judy frame of reference relative to your ground-based frame?"

"Fair enough," Kip replies, "Let's start by taking Judy as our marker. Between spark—at her feet—and click of the photodetector—again at her feet—between the one event and the other, *we* measure her to go forward 8 meters. So *we* must also be the ones to do the timing of that movement. And how? With our trusty clock."

"Clock!" Hans exclaims. "This is the first time I've heard you mention any clock. What clock?"

"Light," Kip explains, "that's our trusty clock. Light—with its standard speed—has only to go a known distance to give us a known time. Five meters of outbound travel on the hypotenuse of one 3:4:5 triangle. Five meters of return travel on the hypotenuse of a second 3:4:5 triangle. Total light-travel distance, 10 meters. Total light-travel time, also 10 meters."

Kip shows Hans our earlier diagram (see page 40) and continues. "In those 10 meters of light-travel time, Judy moved ahead only 8 meters. So she—and the whole Punch-and-Judy frame of reference—are speeding past us at exactly 80 percent of the velocity of light. That's 20 percent slower than light, not 33 percent faster. Agree?"

Hans, finishing his espresso, smiles and nods yes. Hans and Carolee rise and stroll off together. Kip and I decide to meet the next morning to discuss a project, then head off in our separate ways.

Null Interval and Faraway Events

That night, my thoughts shade off into dreams, which now and then come alive as thoughts. How come during the day we hadn't asked another question or two? The start and end of that round-trip flash of light, out from Judy to Punch and back from Punch to Judy, are indeed separated by a 6-meter interval. But what about the simple outbound trip? What is the interval from flash at Judy to mirror at Punch? I sit up and write the numbers on a bedside notepad:

$$(\text{separation in space, spark to mirror})^2 = (3 \text{ meters})^2 = 9 \text{ square meters}$$

$$(\text{separation in time, spark to mirror})^2 = \left(\begin{array}{c} 3 \text{ meters of} \\ \text{light-travel time} \end{array}\right)^2 = \left(\begin{array}{c} 9 \text{ square} \\ \text{meters} \end{array}\right)$$

Subtract!

$$\text{Difference} = (\text{interval})^2 = 0 \text{ square meters}$$
$$\text{Interval itself} = 0 \text{ meters}$$

Suddenly I recognize that the lesson of that one-way flash is the lesson of any photon. The take-off of a quantum of luminous energy establishes one event in spacetime—*here*. Its arrival anywhere, in undeflected flight, establishes a second event in spacetime—*there*. The interval between those two events is always zero. Why? Because the separation of those two events in time, in *light-travel time,* is identical with their separation in space, the distance between those two events. And why this coincidence? Because light has always in every free-float frame of reference the same speed!

On the float, the separation in space between the emission of the flash of light at Judy and its arrival at Punch was 3 meters; the separation in time, also 3 meters; the interval, zero. For Kip and me in our ground frame, their separation in space was 5 meters; in time, likewise 5 meters; the interval again zero. The separations in space and in time are observer-dependent, but interval—the separation in spacetime—is observer-independent. Light and influences propagated at the speed of light make zero-interval linkages between events near and far, tying them together into a richly connected tissue—spacetime, the master of all motions and home of all that was, is, and shall be.

Null interval is a feature of the geometry of spacetime totally different from anything in the geometry of space. Interval, interval, interval! That word dances around in my head as I lie between sleep and wake. It forms itself into a strange new picture of cause and effect as a double cone (see the box on the following page).

My dreaming grows more troubled, as a grinning devil appears. "Old boy, snap out of your fixation. Botheration to your zero interval.

EVENTS AND INTERVALS

The double cone on the facing page reveals the true significance of spacelike and timelike intervals: events separated by a spacelike interval cannot possibly influence each other, whereas those separated by a timelike interval can. Whether one event can influence another event depends on their relative positions in spacetime. A given event (red dot in the diagram) has one of five possible "causal relations" to other events, which are here illustrated as white and black clocks representing successive events in the history of a freely moving object. The table below sets forth the possible causal relations and explains how to interpret the double cone. In the diagram, the regions within the upper cone lie in the future of the red dot: *up* represents later times. The other two dimensions indicate location to the east or west and location to the north or south. The paper is not adequate to represent the third space dimension of four-dimensional spacetime.

The five causal relations between events

When the White or Black clock is	Red dot relative to this White or Black clock represents an event in the	Causal relation between the two events
"Below" lower cone	Future; events separated by timelike interval	Particles as well as radiation emitted from White can affect Red
On lower cone	Future; null interval between events	Only radiation emitted from Black can affect Red
Between lower and upper cone	"Neutral"; events separated by spacelike interval	No effect of Black on Red and no effect of Red on Black
On upper cone	Past; null interval between events	Only radiation emitted from Red can affect White
"Above" upper cone	Past; events separated by timelike interval	Particles as well as radiation emitted from Red can affect White

The upper cone is defined by light rays that spring from the event marked in red. The lower cone is the locus of light rays that arrive at that event. The clocks mark the passage of proper time along a selected worldline. Each clock along the worldline marks an event in spacetime by its tick.

Chapter Three

Some events separated by a 20-lightyear interval

Position of second event	Separation in time (lightyears)	Separation in space (lightyears)	(Time separation)^2 (square lightyears)	(Space separation)^2 (square lightyears)	Difference = (Interval)^2 (square lightyears)	Interval (lightyears)
Here	20	0	400	0	400	20
Satellite	25	15	625	225	400	20
Bellerus	52	48	2704	2304	400	20
Canopus	101	99	10201	9801	400	20

Forget your different ways to get a zero interval. Who cares whether the light flash, as Punch and Judy saw it, took 3 meters of time to go 3 meters of distance? Or whether, as you and Kip saw it, it took 5 meters of time to go 5 meters of distance? Baby business!

"I put to you something really worth talking about, a nonzero interval, a timelike interval of 20 lightyears. Flunk or pass! Tell me the separations in space and time of two events separated by such an interval!" As he grimaces, he presses his pitchfork painfully into my ribs.

"Hold off," I say, hopefully. "One event, here and now. The other event, here and 20 years from now. Interval? Twenty years—or 20 lightyears—whatever unit of time you devils like to use."

"Legalistically correct," he growls, "but too trivial to satisfy me. You've got to do better," he adds, with another thrust of the pitchfork.

I wake up, perspiring. The clock says midnight. I vow to be ready for the next devil that tries to ruin my sleep. After half an hour of figuring, I finish the table on this page.

Back under the covers, a mantralike refrain resounds in my mind as I drop off to sleep, "Canopus, 101 lightyears of time, 99 lightyears of distance, interval 20 lightyears; Canopus, 101, 99, 20; Canopus, 101, 99, 20; Canopus . . . . ZZZ." All through the rest of the night the large star Canopus seems to be flickering in my mind, beckoning me to some adventure.

Planning a Trip to Canopus

Early the next morning, the ring of the phone awakens me. I recognize the voice of a former colleague, Tom Bates, now head of the Space Agency. He explains, with some urgency, that the agency has an important project that he would like me to undertake. I agree to meet him within an hour, wondering what this is all about.

As I put the phone down, Kip arrives. I explain the change in plans and ask him to come with me to the Space Agency. After a quick breakfast, we are off.

Arriving at the Space Agency about 8:30, we walk quickly to Tom's office. Tom wastes no time. "Nearly 99 lightyears away floats the star Canopus, alpha member of the constellation Carinae, The Keel, at a right ascension of 6 degrees. It's doing strange things. We want you to visit it, photograph it, and bring back the records."

"That's impossible," I object. "I have only 40 or so more years to live. I can spare at most 20 years for the outward trip and 20 years for the return trip. Even if I could travel at the speed of light, I would need 99 years merely to get there, let alone come back."

Tom smiles, "I know it sounds impossible, but we know there is a way around that problem. I've set aside an office where I would like you to think it all over carefully. Our planning board wants to meet with you at ten o'clock this morning; and the whole staff, this afternoon. O.K.?"

Tom shows me to a quiet, comfortable office. "I'll be back in about an hour." He wishes me good luck and leaves me to my thoughts.

Too comfortable! I fall asleep in the easy chair. Perplexing thoughts fill my mind. In the midst of dark dreams I hear some cheery notes on the flute. Ariel floats in with a tuneful song. Whatever the words, they induce my lips to murmur my thoughts:

To Canopus and back,
Alive and well,
Only forty years older
Than I am today
Greeting the great,
Great, great, great, great,
Great, great grandchildren
Of my present-day friends,
Two-hundred and two
Years from now.
You are so built, spacetime,
That the indirect route
Through you
Racks up always
Less proper time
Than the direct route.

A few more notes on the flute, and Ariel disappears.

If I let myself go, how much will I really age on the trip to Canopus? Twenty years of aging. Twenty years advance in my biological time.

Twenty years lapse of my wristwatch time. Despite my cheerful dreams, I am dubious.

Ariel peeps in, "Don't forget the magic incantation. *Interval!* You're dreaming about a 20-year interval. And remember that table you worked out so precisely last night."

Suddenly I remember the piece of paper tucked in my pocket. I pull it out and see that miraculously prophetic last line: "Canopus, 101 lightyears of time, 99 lightyears of distance, interval of 20 years."

I awake to a knock on the door. Tom says that it's time to go and escorts me to the planning group. A surge of blood rushes through me. What a miracle that I, of all people, should be invited to undertake a mission so unprecedented! In that instant my decision forms itself, "Yes."

When the preliminaries are concluded and I am called on, I boldly step to the front and present my bare-bones plan: Distance out, 99 lightyears. Time to cover it, 101 years. Speed, $99/101 = 0.9802$ of the speed of light. Interval between leaving Earth and arrival at Canopus, 20 years. Identical figures for the return trip. My aging during mission, 40 years. Aging of mankind on Earth, 202 years.

The group members greet this proposal with enthusiasm and thank me for volunteering for a mission so cosmic. That afternoon I explain the scheme to the entire staff. The great majority welcome it. Al and Bill, however, object.

Relativity: Hoax or Reality?

"Your whole plan depends on relativity," Al stresses, "but relativity is a hoax. You can see for yourself that it is self-contradictory. It says that the laws of physics are the same in all free-float reference frames. Very well, there's your rocket frame, and there's what you call an Earth-linked expanded "room" frame. You tell me that identical clocks, started near the Earth at identical times, time zero, and running—each at rest in its own free-float frame—at identical rates, will read very different times, when they meet again. You claim that in the rocket you will age only 40 years, while I and my descendants—earthbound stay-at-homes—will age 202 years.

"But if there's any justice, if relativity makes any sense," he continues, "it should be equally possible to count you and your rocket as staying home and me on Earth as going away, with all the numbers just the other way around. Nothing could show more conclusively that neither result is right. Aging is aging is aging. It is impossible to live long enough to cover a distance of 99 lightyears. Forget the idea."

"Al," I reply, "we all know that last July you drove straight north on United States Interstate Highway 35 from Laredo, Texas, on the Mexican border, to Duluth, Minnesota, near the Canadian border. Your tires rolled along a length of pavement of 2750 kilometers, and the odometer on your car so indicated. Last fall I also drove from Laredo to Duluth, but I had to make a stop in Cincinnati, Ohio, on the way. I drove northeast as straight as I could from Laredo to Cincinnati, about 2250 kilometers, and northwest as straight as I could from Cincinnati to Duluth, another 1300 kilometers. My tires rolled out a total of 3550 kilometers. Clearly, you can claim that my route deviated from yours, and I can claim with equal justice that yours deviated from mine. That is the relativity of our starts. You took the direct route all the way as conscientiously as I did. The great difference between our travels is this, that my course has a sharp kink in it. That's why my kilometerage is greater than yours by about 30 percent. And there's a like kink in my course through spacetime—or in my *worldline,* if you'll let me use that popular term for a course through spacetime."

Al interrupts, "Are you going to tell me that the kink in the rocket worldline at Canopus is the explanation for the rocket traveler coming back aged less? But that kink makes the worldline longer than 202 years, not shorter."

"Longer in appearance, shorter in reality," I reply, getting up from my seat and heading toward the blackboard. "Let me show you why that's so. On paper, the rocket's journey is depicted as a rightangled triangle whose legs are 99 lightyears of distance and 101 lightyears of time. The slant side, the hypotenuse, of that triangle is 141 lightyears; 282 lightyears for the round trip. So your impression of greater length would be justified.

"In spacetime, however, the slant side of that right triangle is only 20 lightyears. That is the interval—the proper time that the rocket clock racks up on each leg of the trip. During my trip to Canopus, I will experience passage of less time than I would if I remained on Earth. Here's a little table (see page 50) to show that I would come back even younger, relatively, if the Space Agency could come up with a launch velocity still closer to the speed of light. At the maximum speed—the speed of light—I wouldn't age at all, while stay-at-home earthlings would age 198 years."

Al replies, "I still can't believe it is all so simple as you make it out. You must be factoring in some strange effect or hidden time or other deceptive physics at the turn-around point."

I shake my head. "There was no gremlin in the engine pit, nor any other dishonesty at Cincinnati, to make my odometer read 800 kilometers more than yours. A plain no-nonsense feature of Euclidean geome-

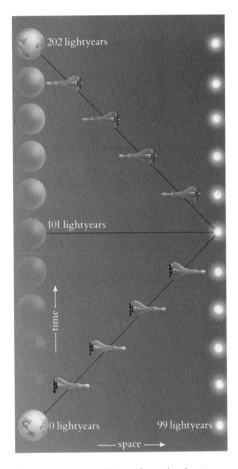

On paper, the worldline of a rocket having a kink in it at Canopus appears longer than that of the straight worldline of stay-at-home earthlings. In terms of proper time as measured by the moving rocket clock, however, the total travel time is only 40 years, much less than the 202-year worldline of the earthlings.

The rocket traveler's aging depends on the rocket speed

The Input	The Details					The Net	The Upshot
Speed out to Canopus and back relative to speed of light	Time in Earth-linked frame for one-way trip (lightyears)	This time squared (square lightyears)	Distance squared (square lightyears)	Difference	Interval itself = aging of rocket traveler going out (lightyears)	Twice this interval = total aging during round trip (years)	Aging of stay-at-home earthlings (years)
0.9801	101	10201	9801	400	20	40	202
0.99	100	10000	9801	199	14	28	200
0.999	99.10	9821	9801	19.6	4.4	8.9	198.2
0.9999	99.01	9803	9801	1.96	1.4	2.8	198.02
1.0	99	9801	9801	0	0	0	198

Note: Years and lightyears are used interchangeably to measure aging.

try gave you the advantage: a straight line makes the total distance between two points the shortest. Spacetime has quite a different geometry: a straight worldline makes the total interval between two events the longest. Here's another drawing (see page 51) to show the difference."

Bill breaks in as Al ponders the table and my two sketches on the blackboard. "All this theory is too much for me. I won't believe a word of this discussion unless you can show me an experiment."

Kip has been listening. He smiles and steps forward. "Bill, what are you really asking? For evidence that a kink in a worldline cuts the aging of a Canopus-bound traveler relative to his earthbound twin? That out and back at 99/101 of the speed of light makes 202 years of earthbound time go by in 40 years of traveler time? That a 40-year worldline with one kink in it can take off at the same place, and end at the same place, as a 202-year stay-at-home worldline?"

Bill is a reasonable man. "I won't believe this relativity business without some rock-hard observational evidence. I'm not asking for an actual experiment with twins or anything like that, but I need something besides words and tables and diagrams to convince me."

"Well, since the journey to Canopus that I am proposing would be a first," I point out, "there never has been a direct test of the 'twin effect' involving people. However, if we trim down the distances, I can show you some laboratory evidence that the 'twin effect' occurs."

"I'm listening," Al replies.

"Let's put aside for the moment the spaceship trip to Canopus. Instead, consider a bullet shot out at the same speed but turned around at a much smaller distance from the Earth. Forty seconds of round-trip time

Chapter Three

will register on each tiny atomic clock placed inside the bullet, but 202 seconds will register here on Earth as the time for the same trip. You agree that the principle is the same?"

"Same principle. Same nonsense," mutters Al.

"After a thousand such round trips," I continue, "202,000 seconds of stay-at-home time will have clicked away, but only 40,000 seconds of bullet time, of bullet aging, of increased bullet seniority. Same principle, right?"

"Yes, again. The same principle, the same nonsense," Al agrees. "But still no evidence!"

"Not so. Just such a multi-zigzag experiment *has* been done, and the time—the seniority, the aging—racked up by the zigzag clocks turned out to be less than the aging or time registered on the stay-at-home clocks—and reliably so, and by an amount in good agreement with the concept of proper time."

"Let's hear the details, then," Al says grudgingly.

"Robert Pound and Glen Rebka, Jr., wound up their wonderful small-scale experimental test of the twin effect in early 1960. Of course, the distance was nothing like 99 lightyears. It was only about one-tenth of the distance between one atom and the next in a piece of iron. And the speed of the zigzagging atom was nowhere near the speed of light. It was only about 300 meters per second going out, and 300 meters per second coming back, the speed of the normal thermal agitation of iron atoms. So you can understand that the seniority, the aging, the proper time registered by an out-and-back iron atom was only slightly less than the proper time registered by a stay-at-home iron atom. The calculated proper time of the kinked worldline for my trip to Canopus and back is only about 20 percent of stay-at-home straight-worldline time. But an iron atom travels at a speed enormously slower than my spaceship. Thus the calculated reduction in seniority for the zigzag route, instead of being 80 percent, as for my Canopus trip, is for a room-temperature iron atom only about one part in a million million."

Al breaks in, "You're telling me that there is a clock accurate enough to measure a difference that small?"

"Yes, an atomic clock," I reply. "It happens that the nucleus inside an atom of mass-number 57 ticks away billions upon billions of times a second and serves as a clock of fantastic precision. Counting those clicks is not the problem in this experiment. The difficulty is that there are no stay-at-home iron atoms in a piece of iron; they are *all* moving. But when a piece of iron is heated, the zigzag speed of the atoms increases; their worldlines kink more sharply. So the atoms in a heated piece of iron rack up fewer clicks in a second of laboratory time than do the ones in a colder piece. And that's not talk. That's fact. Warming iron by just 1°

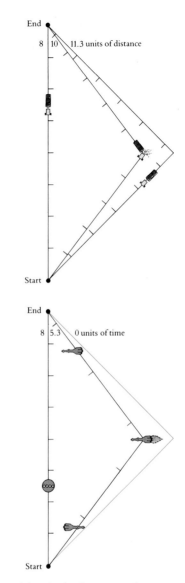

In space (above), the direct route from start to end is the shortest. As the kink in the line of travel becomes more extreme, the distance traveled increases. In spacetime (below), the direct worldline from start to end is the longest (unpowered or free-float spaceship). As the kink in the worldline becomes more extreme, the interval becomes shorter. At the speed of light, the interval is zero; there is no aging.

Atomic-beam clock at the U.S. Bureau of Standards. Cesium atoms boil out of the oven (bottom, right of center) and change direction under the influence of an inhomogeneous magnetic field (bottom, center) that sorts them out by speed. Then the atoms pass through the long tube, where they are twiddled by a weak oscillatory magnetic field. When the frequency of the field coincides with the internal frequency of the cesium atom (9,192,631,770 vibrations per second), atoms resonate, drop out of the beam, and arrive in reduced numbers at the detector. This clock is accurate to one part in 10^{13}.

Celsius cuts down by 7700 the number of clicks ticked off by each internal iron-atom clock during a laboratory second."

"Is that 7700-click difference a large effect?" Al asks.

"No. It's fantastically small compared with the total number of clicks from the internal clock carried by either the cooler or the warmer iron atom, about 3.5 billion billion (3.5×10^{18}). That is a predicted difference of only 2.21 parts in 10^{15}," I explain. "But Pound and Rebka were able to measure that difference. They found that warming iron by 1° Celsius reduced the iron click rate by 2.09 ± 0.24 parts in 10^{15}. That's right on the mark.

"There's your evidence, Bill and Al," I continue. "Real numbers, measured on real instruments in the real world! A marvelous check that the twin effect, relativity itself, is no hoax. It's real. It's part of the working of the world.

"The measurement by Pound and Rebka verifies impressively the idea of spacetime interval. It shows that what counts for the moving internal iron-atom clock is not laboratory time, but proper time, time toted up from start to end along all the segments of the worldline. That's what spacetime geometry is all about.

"Seniority or aging or proper time, as measured by a carried-along clock is what counts for a moving object, not laboratory time," I summarize. "This lesson stands out in experiments of the greatest variety. In some, the two time measures differ enormously more than they do in the Pound-Rebka experiment. In some, the precision of the time measurements rivals their work. In no other experiment is the connection clearer to the chronometric compression caused by the kinked course of the coming carom on Canopus."

Kip breaks in, "John is right. The iron atom goes back and forth at a speed high by automobile standards but low compared with the speed of light. That is the only reason for the smallness of the kink effect in the Pound-Rebka experiment. About the reality of the shorter time, there can be no question." Looking at Al and Bill, he asks impatiently, "Do you two gentlemen now agree that the Canopus trip *is* feasible?"

Al and Bill confer briefly and then chorus in unison, "We concede."

Without further dissent the entire staff votes to accept my plan. Canopus here I come!

All of Space Is Ours

Can we mortals ever reach the speed of light? No. However, to harp always on that *no* is to lose sight of the wonderful *yes* message of spacetime geometry. Yes, in one year of our own proper time, we can—

traveling in a sufficiently fast rocket—penetrate any number of lightyears of space, go anywhere we choose. Do we want to go to the most distant known star and get there in our lifetime? Then, granted a sufficiently fast rocket, we can do it. With speed close enough to the speed of light, we can make the aging factor so small that while the Earth and stars age 50 billion years, we age 40 years. The extent of space and time that in principle is accessible to a single mortal is almost inconceivable. Despite the enormous practical problems, space colonization is changing from dream to design, and soon to doing. Billions of lightyears of space wait for life to animate them. Billions of years lie ahead to capture the billions upon billions of sites useful for life. The universe will grow ever more exciting. That's reason enough to want to stick around and drop in on act after act of the show. We someday can!

Not one word of spacetime curvature in all this. All is straightforward consequence of the physics in a free-float frame of reference: event, and the spacetime interval between two events. *The interval is the interval is the interval.* The interval between any two events has an existence and value independent of the free-float system of printout clocks by which it is registered, independent even of any registration at all.

Events and the interval between events build spacetime. This is the marvelous and curious but wonderful way that the world is put together. The interval is the building strut of flat spacetime, the local spacetime of a local free-float frame. It is also the building strut of curved spacetime, spacetime in a region where gravity is strong enough to make itself felt.

4

Boomeranging through the Earth

Spacetime, are you endowed
With ideally flat
Mathematical perfection?
Yet do you,
Through your grip on mass
Rule all motion?
What a disaster
That combination would be!
Two masses, starting at the same point
In directions differing in the slightest
Would go on in their straight lines
Separating ever more
With each passing year.
Earth, Sun and all we hold dear,
Would fall apart.
You are curved!
Your curvature
Reveals itself to us,
Clear and forcible,
In gradual changes
In the relative velocity
Of two nearby travelers
Boomeranging
Through an ideal Earth.

The double boomerang shaft through Earth's center from Taralga, near Sidney, to the Atlantis Seamount.

Like a perpetual-motion boomerang, an imaginary boomeranger would travel in free float back and forth from one side of the Earth to the other.

The idea is old of a shaft drilled through the center of the Earth, a direct route from west to east. The brave traveler releases his fingerhold on the western rim of the terrifying hole and 42 minutes later he coasts to a safe landing at the eastern rim on the opposite side of the Earth. Is there a variant of this story in which the traveler fails to grasp the eastern rim and ends up, after another 42 minutes, back at his starting point, to win from his joking friends the title of "Boomeranger"? If so, I never heard of it, nor of the use of the word *boomeranging* in this connection. Nonetheless, let's adopt that term right now for this imaginary form of travel.

Few examples of gravity at work are easier to understand in Newton's terms than boomeranging. Nor do I know any easier doorway to Einstein's concept of gravity as manifestation of spacetime curvature. Nowhere better than in a close look at boomeranging are we able to compare and contrast the two ways of understanding gravity. The direct flight of an imaginary earthship through the Earth is much simpler to describe than the travel of a satellite in a circular orbit around the Earth. The satellite needs two dimensions of space to tell where it is and the one dimension of time to tell when it's there. Any curvature of that three-dimensional spacetime is hard to visualize and harder still to depict on paper. In contrast, we need only one dimension for space and one dimension for time to describe the motion of a spacecraft through the Earth. They give us a two-dimensional spacetime, whose curvature will be much easier to illustrate on paper.

Planning the Boomerang Project

Boomeranging. No better word is there to describe the imaginary journey of one who falls through the Earth with never an obstacle to bang into, and so comes back to his starting point. As we begin planning our Boomerang Project, we soon realize that an enterprise so vast, so novel—even if only in the imagination—requires the very best minds. We quickly assemble the Boomerang Project Organizing Committee consisting of experts from around the world. We explain the objectives of the project to the committee and why we have named it the Boomerang Project. In their first decision, the committee members vote unanimously that to honor the Land of the Boomerang, one end of the shaft should be located in Australia.

Getting down to their deliberations, the committee members discover that lines running straight through the center of the Earth from various points on the coastline of Australia stake out an area filled with heaving North Atlantic waves, with not one bit of land anywhere in it. Consternation! No place for the other end of the shaft to exit—at least not above water. Imagine the relief of the committee members at finding in all that watery waste one point of promise, a point on the Midatlantic Ridge called the Atlantis Seamount. Because this point lies only 82 fathoms—about 150 meters—below the surface, the committee agrees that a watertight chamber, a caisson, can be built from the level plain of the seamount to the ocean's surface, thus creating an artificial island. By day's end, the other details of the project are worked out. The committee then adjourns, after setting a starting date for construction.

The Boomerang Project begins! The caisson is built, the water is pumped out, and the proud flag of the new principality of Atlantis is planted by committee members at its highest point. Next, with the most modern machinery available, drilling of the shaft begins simultaneously from Atlantis and from its antipodal point, the little town of Taralga in New South Wales. A superstrong lining is installed in the shaft to prevent a cave-in under the enormous pressures deep in the Earth's interior; insulation also is installed to protect against the great temperatures there.

Crisis! One of the committee members, checking on the progress of the lengthening shaft, suddenly realizes that any craft that journeys through the shaft will scrape against its sides because the Earth is rotating. At a hastily called meeting, the committee decides on an extraterrestial solution to this problem: two giant hands reach down from space to still the Earth's rotation, and then those hands use a giant bicycle pump to inflate the denser-because-slumped center of the Earth. The desired result is achieved: the entire planet is nonrotating and of uniform density. There is great excitement all over the Earth as the two drilling

Rob and Alix discuss the Boomerang
Project with the Australian ambassador.

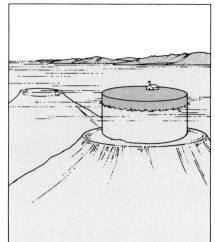

One end of the boomerang shaft exits
at the Atlantis Seamount.

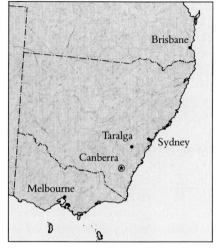

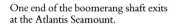

Taralga in New South Wales is
the eastern exit point of the
boomerang shaft.

teams meet at the center of the Earth. They christen the spot Hyperion—
that unique place in all the world where one can experience weightlessness without having to fall to do so. The drilling teams return to the
surface, seal both ends of the shaft, and pump it out to a high vacuum, so
that earthships can boomerang through it without resistance and without
being fried to a crisp.

With completion of the boomerang shaft, the committee's job is
done. We thank them for their help, noting that the first trip through the
shaft will start in one week, starting from the Atlantis end.

The Maiden Boomerang Voyage

All is ready. The maiden voyage of the earthship—dubbed Boomer I—is
about to begin with Rob at the controls. We brief him on what to expect.
He and his earthship will zoom effortlessly back and forth through the
darkness of the evacuated boomerang shaft, Atlantis to Taralga, Taralga
to Atlantis, Atlantis to Taralga . . . . Mile after mile, hour after hour, he
and Boomer I will float free. He will see no weight, feel no weight, have
no weight. If he bounces a pingpong ball against the wall of the brightly
lit craft, its path will reveal to him no fall, no curve, no sign of gravity.
Spacetime, master of motion, will appear to be as flat here as it is in the
darkness of space far from planet, sun, or galaxy.

To prevent the boomerangers from scraping the wall of the tunnel, the Earth's rotation must be stopped.

Inflating the inside of the Earth ensures uniform density.

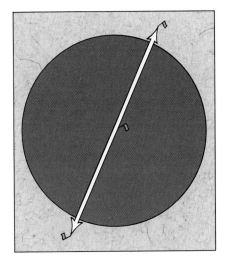

The boomerang shaft is completed.

In dismay we realize that Rob's boomeranging will not show us anything about spacetime curvature. Is the whole Boomerang Project a bust? No, I assure everyone. We must look at the motion of two nearby boomerangers relative to each other, rather than at the motion of a single earthship relative to the Earth. In this way we can focus on local physics, as Einstein advises, not action-at-a-distance physics, Newton's original doorway to gravity. The wisdom of the organizing committee in planning a double boomerang shaft, like the tunnel of a major highway under a river, is now obvious. Not for counter-current traffic, however, do we need the second Atlantis-Taralga tube, but for a second earthship—Boomer II. Alix eagerly volunteers to pilot this craft, which will start off from Atlantis 2 seconds later than Rob does in BI. In consequence, Alix and BII initially will be 20 meters behind Rob and BI.

We've designed both earthcraft to be too light in mass to influence one another. They respond to the gravity of the far larger mass of the Earth without detectably influencing it or each other. In this sense they serve as *test masses*.

As soon as the pre-boomerang briefing is completed, we all assemble at the Atlantis passenger platform. At exactly 1 second before 11:18 A.M., Rob frees BI from its mooring and begins zooming past feature after feature of the Earth's interior. Alix departs 2 seconds later and passes the same landmarks, always 2 seconds after Rob does.

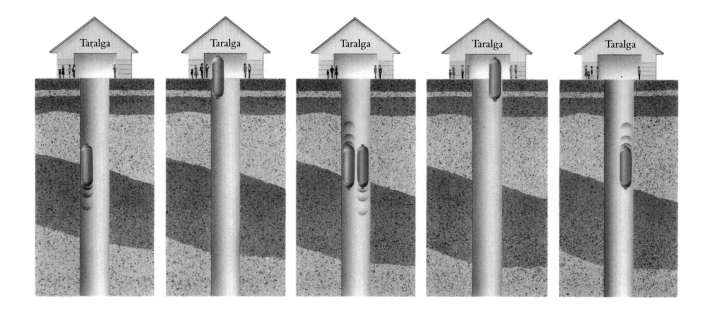

Rob and Alix approach the Taralga platform for the first time 42 minutes after their departures. Their paths cross at noon, Rob now descending and Alix still rising but slowing down. Their 42-minute minuet will continue until they moor at one of the platforms.

One second before noon Rob is at rest for a fleeting instant at the level of the Taralga passenger platform. Without a jolt he could then and there lock his craft to the wall of the shaft if he so choses. He doesn't. He wants to keep on longer in free-float travel. At this instant Alix is 20 meters below the platform and still rising fast.

She finds her rise to be half as fast at noon itself. Alix is in the final slowdown of her own approach to Taralga. At this point, only 5 meters below the platform, rising Alix meets descending Rob. He has already covered the first 5 meters of his return trip to Atlantis.

One second past noon Alix is at the summit of her flight. Rob is descending at twice the speed he had a second ago. He is already 20 meters down.

Two seconds past noon Alix herself is 5 meters down, well started on her return trip to Atlantis.

A minuet is in progress. The two boomerangers execute its next movement at 12:42, near Atlantis. There they encounter each other again; Alix slowing as she rises, Rob gaining speed as he starts downward on a new trip to Taralga.

The passage of yet another 42 minutes finds them passing once more, near Taralga.

The minuet goes on and on, ever maintaining its 42-minute rhythm, even as the two masses, in free float, zoom through the center of the Earth at the fantastic speed of 7.9 kilometers a second or 18,000 miles an hour.

Curvature of Spacetime in the Boomerang Shaft

By late afternoon, it is clear that the Boomerang Project is a great success. Alix and Rob are fine. The earthships have survived intact. On their next approach to the Atlantis platform, the two boomerangers lock into the moorings and disembark, to the cheers of all, from BI and BII.

The amazement of it! We have seen two test masses, both in free float. They were together at high noon, at the point where their paths cross 5 meters below Taralga. As they left the crossing point, they floated apart in the darkness of the shaft. The separation between them every second grew 20 meters greater. Imagine Rob and Alix thus to be parting company in some faraway region of the cosmos where spacetime is essentially flat. Straight as arrows, their tracks would carry them to ever greater remoteness. Goodbye forever! Within and near the Earth, however, spacetime is not flat, nor is it so within the Atlantis-Taralga shaft. Not forever do Alix and Rob continue to increase their separation by an additional 20 meters with each passing second. As the metronome of the minuet prepared to boom out the 42-minute beat, Alix and Rob were both coming up on Atlantis. What is more, Alix was shortening Rob's lead. Each second then, the distance between them was not 20 meters greater, but 20 meters less!

The Alix-Rob separation increased fast in the first phase of the Taralga-to-Atlantis trip. It decreased fast in the final phase of that journey. Clearly at some place between the two terminals a changeover occurs from slow increase to slow decrease in separation, a place where that separation isn't changing at all—not changing, because it has crested. Crested where? Obviously at the center of the Earth. Crested at a separation how great? Sixteen kilometers, according to a bit of figuring that we can skip.

Sixteen kilometers. Roughly only a thousandth of the diameter of the Earth. A peanut distance! Thus we're talking about *local* physics when we're analyzing the Alix-Rob separation. And Einstein tells us that physics only looks simple when we analyze it in local terms. So no more mention of Atlantis, or Taralga, or the position of Alix and Rob relative to the Earth. Only the distance between the two earthships is relevant.

BOOMERANGING ACCORDING TO NEWTON

Not Einsteinian *free-float* through a curved spacetime, but *force* acting at a distance through flat geometry—that is in brief the difference between today's account of boomeranging and Newton's. According to Newton, every bit of matter in the universe attracts every other bit of matter with a force proportional to the product of the two masses and inversely proportional to the square of the distance between them. This is enough to tell us that an earthship will oscillate back and forth in the boomerang shaft through the ideal Earth, making one complete cycle every 84 minutes in perfect synchrony with a satellite going round and round just above the surface of the Earth (the friction of the air being somehow evaded!)

In Newtonian terms, boomeranging is analogous to the back-and-forth oscillation of a pendulum, or the up-and-down vibration of a weight hanging from a spring, or the singing of one of the arms of a tuning fork that has just been struck. The key points are (1) a mass, (2) an equilibrium point for this mass, (3) a displacement from this equilibrium point, and (4) a restoring force proportional to this displacement. These requirements granted, the mass oscillates. For the mass to complete one complete round trip requires one complete period: 84 minutes for the boomeranger, less for the other oscillations. This period, morever, is independent of the magnitude of the motion, large or small, provided only that the restoring force really does increase in direct proportion to the distance from equilibrium: double for double the distance; threefold for three times the distance; five times greater for a fivefold greater distance.

A five-times-greater force on the boomerang craft when it is five times farther from the center of the Earth. How can this be when Newton interprets gravity as a force between every mass in the universe and every other mass that is inversely proportional to the square of the distance? Such a force, at five times the distance, is not five times greater, but 25 times weaker. To resolve this apparent contradiction, Newton divided the attracting mass, in imagination, into two parts that are separated from each other by an invisible boundary in the form of a sphere centered on the center of the Earth and passing through the test mass. The mass outside that imaginary boundary constitutes a hollow sphere. The different parts of that hollow sphere pull on the test mass with different strengths and in different directions. The net resultant of these forces, Newton proved from his inverse-square law of gravity, is zero, no pull at all. Only the mass inside the imaginary boundary exerts any net attraction. That mass pulls on the test mass with the same strength as if it were all concentrated in a single point at the center.

Newton's reasoning lets us understand why, in the boomerang shaft through an Earth of ideally uniform density, the pull of gravity toward the center is exactly proportional to the distance from the center. Consider, for example, a point one-fifth of the way from the Earth's center to its surface. Draw in Newton's imaginary sphere, centered at the middle of the Earth and passing through the earthship. This sphere is smaller than the great sphere of the Earth in its length, width, and breadth, that is, in all three dimensions. Thus its volume is only one-fifth of one-fifth of one-fifth of the volume of the identically shaped Earth. It contains only $(\frac{1}{5})^3$ of the number of cubic meters, only $(\frac{1}{5})^3$ of the number of metric tons, only $(\frac{1}{5})^3$, or $\frac{1}{125}$, of the mass of the Earth. However, the inverse-square law of gravity being what it is, an earthship experiences from each of these fewer metric tons of mass—regarded as being focused at a single point, the center of the Earth—a pull 25 times greater than it would experience at the fivefold greater distance of the Earth's surface. The conclusion is easy to state. A mass exerts a net force on a test mass that is proportional to the number of tons of mass multiplied by the pull exerted by each ton. Alternatively, the force is proportional to the cube of the distance from center to test mass divided by the square of the distance from center to test mass. In brief, the force on a test mass is directly proportional to its distance from the center.

To explain the oscillation of a boomeranging earthship, Newton's theory envisages an imaginary sphere whose outer boundary passes through the earthship. Only the mass inside the sphere exerts any pull on the earthship.

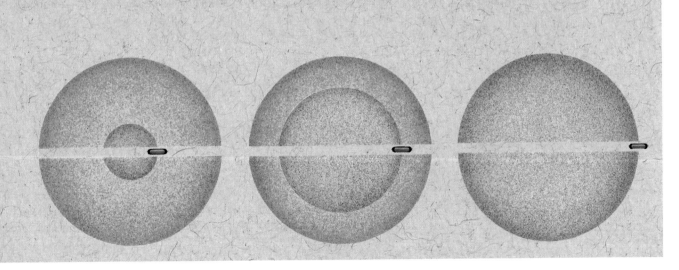

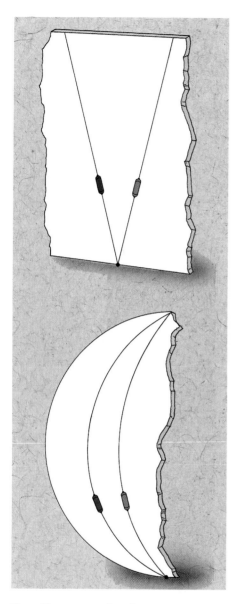

Top: *If space were flat, then two boomerangers, or any two moving test masses, would move apart forever.* Bottom: *Because of the curvature of spacetime locally, where the test masses are, this does not, in fact, happen.*

What can we learn from the Alix-Rob separation about the nature of spacetime in a local region? That spacetime geometry does not let two local test masses fly apart forever. Initially, to be sure, Rob and Alix each saw the motion of the other as a uniform speed of separation, a constant recession velocity. When their separation became larger, however, the recession velocity began to diminish from minute to minute. This slowing of the recession velocity takes place faster, the larger the separation itself is. In other words, the decrease in the recession velocity in one second is directly proportional to the average value of the separation itself during that second.

What happens when there is little or no actual separation, but a substantial speed of separation? Then little or no change in the recession velocity occurs in the course of a second. What happens when there is a larger separation, hundreds of meters, or even some kilometers? Then in the course of one second, the recession velocity decreases quite measurably. This decrease in the speed of separation depends only on the average value of the separation itself in that second, not at all on what the speed of recession happens to be.

As Alix and Rob boomeranged through the shaft, their separation, even at its greatest, was never more than a tiny fraction of the Earth's diameter. What brought free-float Alix and free-float Rob back together again time after time, when they would have separated forever if they were traveling through the flatness of faraway space? Ask this question as if freshly arrived in this magnificent universe. Cast off all school memories of a force acting at a distance, all misleading Newtonian overtones that go with the old word *gravity,* all dreams of some monstrous green puller hidden in the cracks and crannies of space. Instead, look for an answer, a local answer, a believable answer, at space and time right where the boomerangers are.

Spacetime geometry! Why invent anything more? The two earthships plough ahead free-float on the straightest world lines they possibly can. Every step of the way they acknowledge as guide and master the geometry of space right where they are and nothing more. But that geometry cannot be flat; otherwise Alix and Rob would separate forever. It must be curved. Only so can we understand in spacetime-geometry language why two worldlines, originally separating, and both conscientiously ploughing straight ahead nevertheless slowly turn, come back together, and cross. It is not the worldlines themselves that curve; it is the spacetime in which the earthships travel in free float that curves.

In addition to evidence of this spacetime curvature, we get a direct measure of the curvature from the motion of our two boomerangers. The ratio between the change in their speed of separation and their separation supplies a clean, clear measure of spacetime curvature right where the boomerangers are:

$$\text{spacetime curvature} = \frac{\text{decrease in separation speed per unit time}}{\text{separation}}$$

Because the initial separation between Rob and Alix was small, only 20 meters, their initial recession velocity of 20 meters per second decreased slowly. A second later, when the separation had increased to 40 meters, the separation itself was still increasing. However, the speed of separation was then diminishing twice as fast as it was an instant before. When the separation between Rob and Alix approached its peak value, 16 kilometers, the *speed* of separation between the two boomerangers dropped at its fastest pace. It dropped to zero, then dropped to negative— that is, changed from speed of separation to speed of approach.

Three very different separations, and three very different rates of change in separation speed. Amidst these differences, however, one pillar of constancy stands out: the ratio of the change in recession velocity to separation always has the same value. We will work out the value of this ratio in the next chapter. Suffice it to say here that the value is small, only 1.73×10^{-23} per square meter. So small because the spacetime curvature imposed by matter as "dilute" as the material of the Earth *is* small!

Spacetime curvature! That's gravity in brief. And we have a picture for it. More than a picture, we have a number to measure it, a number by which to grab that curvature up for honest examination, a number to prove that we're talking about something more than hot air. But what can we say about the cause of curvature? That question leads us from the action of curvature on nearby test masses, which the Boomerang Project illustrated, to the action of the Earth's mass in generating curvature.

5

Tides: The Grip of Mass on Spacetime

Oh Spacetime, child that you are of time
And of the innocent emptiness of space,
You are both stage and actor
In the great drama
Of all that was and is
And ever shall be.

Two free-float masses
You guide together or apart,
According as your curvature is positive or negative.

Outside the Earth
You display a tide-driving curvature.
There, when you act on a free-float,
Initially static spherical cluster of test masses,
You squeeze it in two directions,
Stretch it in the third,
Give it the shape of an egg,
Unchanged in volume.
Totally noncontractile,
Totally tide-driving,
You thus show yourself outside the Earth.

Inside the Earth, you display contractile curvature,
Witness by its magnitude in any region
To the density of mass in that very region.

Mass thus grips you, oh Spacetime,
And wherever it is
Forces on you there a curvature
Not tide-driving, but contractile.
By this response of yours to mass,
You reveal the Grip of Gravity.

The Moon's tidal drive strips miles upon miles of tidal estuary to its brown bottom.

Gravity. We most commonly observe its effect as the fall of a test mass toward a big mass—an earth, or moon or sun, a faraway mass. But haven't we given up this action-at-a-distance, this Newtonian, view of gravity? How then do we get from this view, so consistent with our everyday observation—to Einstein's view of gravity as the effect of spacetime curvature on the relative motion of nearby free-float masses?

How does Einstein's local-only action explain Newton's faraway action? Briefly stated, mass *there* bends *local* spacetime there. This faraway spacetime forces a slightly lesser bending on the *local* spacetime immediately around it—even though this space is free of mass. This curvature, still faraway, in turn imparts a curvature—a still smaller one—to the *local* space still farther out from the mass. And so on, all the way out to the *here* where we are and where we see "gravity" showing itself. As Einstein describes gravity, all the way from the mass there to us here, every action is a *local* action.

Then shall we throw away Newton's idea? Give up the view that every mass in the universe pulls on every other mass with a force of attraction proportional to the product of the two masses and inversely proportional to the square of the distance between them? No, keep it in the closet. Sometimes—for high speeds, or fast-wiggling masses, or gravity waves, or approach to gravitational crunch—Newton's view doesn't work. But keep it handy as a crutch, a crutch to help us make our way to Einstein's battle-tested, experiment-tested view of gravity.

Curvature: The Grammar of Gravity

Many of the features of curvature show themselves as clearly on the two-dimensional space of the surface of a potato as they do in the two-dimensional spacetime of the two boomerangers we followed in the last chapter. The coordinates of that spacetime are the Rob-Alix distance and the time registered on Rob's wristwatch. And what about *four*-dimensional spacetime, the medium in which we and the Earth, Moon, and stars exist and move? We can see, understand, and appreciate much about its full-blown curvature from a little tour through the land of two dimensions.

Of all the features of curvature, the one most evident is the difference between positive curvature and negative curvature. Examples of each abound, but nothing shows positive curvature more simply than the surface of a basketball; nothing shows negative curvature more simply than the surface of a saddle; nothing shows zero curvature—flatness—more simply than a sheet of paper.

On the surface of a piece of paper, two nearby ideal lines that start

Chapter Five

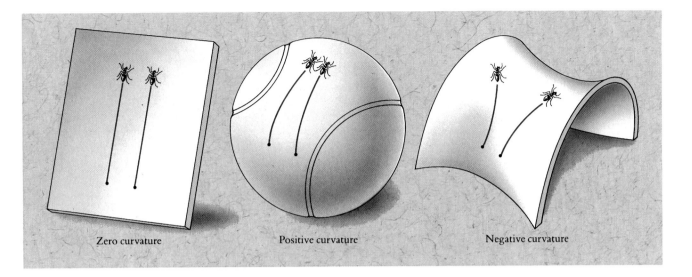

| Zero curvature | Positive curvature | Negative curvature |

out straight and parallel keep on forever at the same separation. So do two ants that travel along those lines. On the surface of a basketball, two nearby ideal lines that start out straight and parallel begin to *converge*. So do two ants that travel along those lines. On the surface of a saddle, two nearby ideal lines that start out straight and parallel begin to *diverge*. So do two ants that travel along those lines.

What brings together two ants that pursue their courses on a basketball? We don't call it gravity. We call it curvature, positive curvature, positive *space* curvature. What brings two boomerangers together, again and again, in their free-float existence in the spacetime interior to the Earth, a spacetime warped by the presence of the Earth? We call it positive *spacetime* curvature. We saw evidence of that positive spacetime curvature in the minuet of the nearby worldlines of Alix and Rob as they boomeranged through the Earth, their courses crossing over and over again every 42 minutes. The spacetime through which those two free-float travelers made their way has positive curvature, curvature like that of the surface of a sphere. This we discover not only from the long-term evidence, the interweaving of their tracks through spacetime, but even more immediately from the short-term evidence, the bending of their two worldlines toward each other.

We cannot discover the curvature of the spacetime that pervades the Earth from the worldline of Alix as she zoomed through our planet. Why not? Because her earthship followed through spacetime the straightest possible course, an ideal track, a free-float history. Nor can we discover the curvature of the spacetime inside the Earth from Rob's free-float worldline, and for the same reason. Not the one worldline and

Initially "parallel" lines remain equally spaced on a flat surface with no curvature, converge on a surface with positive curvature, and diverge on a surface with negative curvature.

not the other, but the local bending of the one worldline toward the other is the signal of the curvature of spacetime.

As Rob and Alix zoomed through the center of the Earth, each pursued—as always—a free-float worldline. At the center their two worldlines reached their maximum separation, 16 kilometers. Momentarily, the two earthships had stopped receding from each other and had not yet begun to converge; thus their two worldlines, separated by 16 kilometers of space, were tracing out parallel courses through spacetime. Nevertheless, they began to approach, ultimately to cross. The two worldlines thus manifested relative bending—bending not of either worldline individually but bending of the course of the one as seen and registered by its distance from the course of the other. This change in the speed of approach from zero to a positive value signals positive spacetime curvature.

The bending of one free-float worldline toward (or away from) a similarly directed nearby free-float worldline gives not only *local* evidence of spacetime curvature but also *local* evidence of gravity. Gravity is a *local* something, and this *local* something is spacetime curvature. No view of gravity is in the end more powerful. No view of gravity is in the beginning more puzzling.

Who that puzzles about "curvature of a geometry" cares whether that geometry is the curved two-dimensional surface of a potato? Or the curved three-dimensional geometry of the space around Earth at this very instant? Or the curved four-dimensional geometry of the spacetime that surrounds Earth from the past through the present on into the future? Whatever the dimensionality—two, three, or more—the initial difficulty is the same: the apparent vagueness of the idea.

A Free-Float Worldline Is a "Geodesic"

So let's start with the potato, a huge one, with many a lump, wart, and bump—ugly, perhaps, but easy to get, to handle, and to study. On its surface our pen, or an ant, or a centipede, traces out a line. That line is most definitely curved, but is it also straight? That is the question with which to begin. George Orwell's famous book *Animal Farm* tells of a world where everyone is equal, but some are more equal than others. The potato—or a curved geometry of any dimensionality—shows us a world where every line is curved, but some are more curved than others.

The conscientious centipede ploughs straight ahead, as straight as it possibly can, deviating never the slightest either to left or to right. The line it follows we cannot call "straight." A clear distinction exists, how-

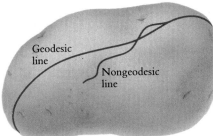

On the curved surface of a potato, both a geodesic and nongeodesic line are curved, but only the geodesic line is also straight, that is, wanders neither to the left nor right along its entire length.

ever, between such a line and other lines: lines that have no such compelling sense of direction, lines that meander, lines that veer off to the left or right. The conscientious line for more than a century has held the standard title, "*geodesic.*"

No deviation from the direct onward course—that is one way to distinguish a geodesic from other lines. There is another. The geodesic has *extremal* length. That is, of all the lines in space that connect a specified starting point with a specified (and not too distant) point of termination, the geodesic line is the one for which the length is *least*. And in spacetime? Of all the timelike worldlines—some wiggly, some smooth—that connect a specified initial event with a specified later event, the geodesic is the one for which the proper time elapsed, the cumulative stopwatch reading, the aging, is the *greatest* (see illustration on page 72). The geodesic worldline is the force-free history of travel from one event to another. It is the free-float worldline.

In the absence of a conscientious centipede, we can mark out a geodesic in the two-dimensional space of a potato's surface with a plastic straightedge—thin, narrow, flexible, and transparent. One end of this straightedge we put down on the potato at any point, pointing in whatever direction we choose. No freedom does it have to wander off to left or right as we push it down, centimeter by centimeter, into contact with the potato. The geodesic that it gives we mark down step by step on the potato skin before releasing our hold on the straightedge. Nearby we inscribe a second geodesic on the surface. Where the two geodesics on a given region of the surface bend toward each other, the curvature of that region is positive; where they bend away from each other, the curvature is negative. Out of this relative bending of nearby and nearly parallel geodesics we will later see how to define and measure curvature.

In spacetime we need no manufactured ruler, no conscientious application of that ruler to a surface, in order to construct geodesics. Nature does it for us. The free-float worldlines of test masses like Rob and Alix are each geodesic. Each pursues the most direct possible course through spacetime. The curvature of that spacetime reveals itself by the bending of nearby worldlines relative to each other.

Of this power of spacetime over mass, Isaac Newton surely had a premonition when, finishing his great book on gravity, he penned into his brief preface these two sentences on the 8th of May 1686: *Nam & linierum rectarum & circularum descriptiones, in quibus geometria fundatur, ad mechanicam pertinent. Has lineas describere geometria non docet, sed postulat.* (The tracing out of both straight lines and circles, on which geometry is founded, belongs to mechanics. How to trace out those lines geometry does not teach, but postulates.) In brief, the science of geometry de-

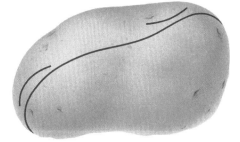

Where two nearby geodesics on the surface of a potato bend toward each other, there the surface has positive curvature; where they bend away from each other, there the surface has negative curvature.

In the spacetime of a trip to Canopus and back, the straight, nonkinked worldline—the geodesic worldline—racks up the greatest aging, or proper time. In the extreme a worldline that goes out with the speed of light and returns with that speed has maximal kink—it does not age at all; it has a null interval between its departure from and return to Earth.

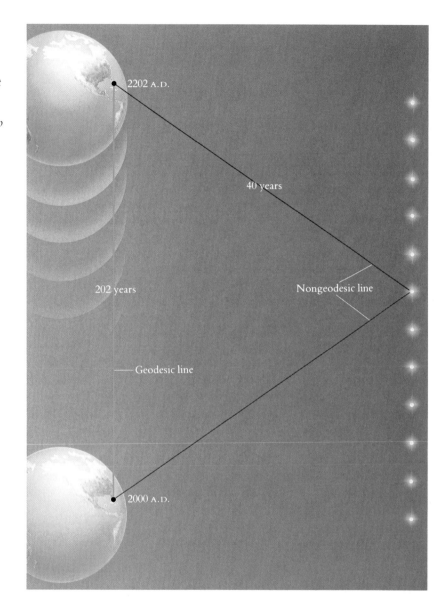

2202 A.D.

40 years

202 years

Nongeodesic line

Geodesic line

2000 A.D.

mands geodesics, but only nature—yesterday in the form of Newton's mechanics, today in the grip of Einstein's spacetime on mass—supplies them.

Distance along a geodesic, distance between two nearby and nearly parallel geodesics, distance between nearby points: all these ideas are reasonable and natural to use in coming to terms with curvature of a two-dimensional geometry, whether it's the surface of a ball, the surface of a saddle, or the spacetime of boomerangers. However, in all our thinking about surface curvature, let us here and now abjure any appeal whatsoever to the most obvious feature of ball and saddle. To be sure, we ordinarily see and think of these surfaces as embedded in our nearly flat everyday three-dimensional space. But no longer! Or, at any rate, not here! Here we seek to employ only concepts that belong to the world of two dimensions. A curved two-dimensional geometry, yes; but not one feature of that geometry makes the slightest call on any third dimension to express, picture, or define its curvature. Likewise when we turn to the full four-dimensional spacetime that is stage for—and participator in—the great drama of existence. We can best understand its geometry and its curvature when we totally rule out any thought of embedding that geometry and its curvature in some flat geometry of some higher dimensionality.

Marvel that is no marvel: we can understand, define, and measure the curvature of a geometry by way of distances in space and intervals in spacetime, both of which show themselves entirely within that geometry. We do not need the flatness of a higher dimensionality in which to embed a curved-space geometry. Openheartedly we can accept curvature for itself; accept it with all the warmth with which we enter on any other newly gained aptitude—skiing, swimming, hang-gliding.

Curvature is positive when nearby and nearly like-directed free-float worldlines bend toward each other. Curvature is negative when they bend away from each other. Curvature is zero when they stay parallel or maintain a constant speed of recession or approach. But surely there are degrees of curvature? More curvature here, near the Earth; less curvature there, far from the Earth. Yes. Let's see then how we can measure curvature.

Measuring Curvature

To measure curvature of spacetime is to assign to each elementary portion of spacetime area its proper responsibility for geodesic bending. Every term in this concept lends itself to illustration and easy measurement: bending, spacetime area, and assignment of responsibility. Of all places to see these ideas in action, none presents itself to us more immediately than the adventure of the two boomerangers, Rob and Alix.

Space—the separation in space between Rob and Alix.

Time—the change in their speed of approach, as measured by Alix and Rob themselves, at one instant and a given *time* shortly thereafter.

Bending—the sharpening in the tilt of their worldlines relative to each other as reflected in the change in their relative velocity.

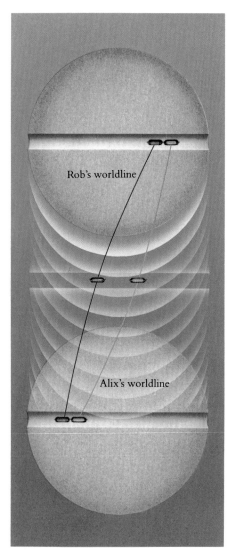

The spacetime area responsible for bending of the boomerangers' worldlines relative to each other—from 1 second before to 1 second after attaining their maximum separation—is bounded by Rob's worldline (red), Alix's worldline (blue; her tunnel hidden behind Rob's), and the lines connecting their positions in space at the two times. (The diagram is not drawn to scale.)

Responsibility for this bending—the product of two factors. One is their separation in space. The other factor is the separation in time between the two measurements of approach speed. Space—the space between Rob and Alix—is not alone what counts. Time—the time between the two measurements of approach speed—is not alone the determining factor. The change in rate of approach between one moment and the next is proportional to the product of the space factor and the time factor, that is, proportional to the spacetime area embraced by the two measurements.

The spacetime area embraced is bounded by four lines. On one side is the worldline of Alix. On the other is the worldline of Rob. Ahead lies the later time boundary. This line runs through space, connecting the positions of Alix and Rob at the instant of their second measurement of approach speed. Behind lies the earlier time boundary. It too is a line through space. It too connects Alix and Rob—but at the instant of their first measurement of approach speed. The spacetime area embraced by the four boundary lines (Alix, Rob, and the times of the first and second measurements of approach speed) has an area as surely as does a building lot. To that parcel of land we attribute an area, in square meters, given by the product of its width and its length, each expressed in meters. Equally well bounded, equally well defined, equally well expressed in square meters is the spacetime area responsible for the change in the Alix-Rob approach speed.

Good. Then attribute to each square meter of that spacetime area its own due share of the change in approach speed. That share, that local bending power, that per-square-meter responsibility for changing the approach velocity, is *curvature*:

$$\text{spacetime curvature} = \frac{\text{change in approach speed}}{\text{spacetime area responsible for that change}}$$

In a given region of spacetime, this ratio between change in approach speed and spacetime area is a constant, independent of the actual values of the change in speed and spacetime area.

We can actually calculate this curvature from what happened to the Alix-Rob approach speed between 1 second before they attained maximum separation and 1 second after. First we look at the change in approach speed. Then we consider the amount of spacetime area responsible. Finally we simply divide the one by the other to get the amount of bending assignable to each square meter embraced; that is, the curvature.

At 1 second after attaining their maximum separation at the center of the Earth, Rob and Alix were approaching each other at a speed of 2.4

Chapter Five

centimeters a second, or 0.024 meters per second. But wait, we do not want to sneak that nongeometric measure of time, the second, into a purely geometric description of gravity. The solution for this little problem? Express the speed of approach in units of—that is, relative to—the speed of light:

$$\frac{\text{speed of approach in normal time units}}{\text{speed of light}} = \frac{2.4 \times 10^{-2} \text{ meters per second}}{3 \times 10^8 \text{ meters per second}}$$

$$= 0.8 \times 10^{-10}$$

Thus, the speed of approach is 0.8×10^{-10} of the speed of light. And the speed of approach 1 second before Alix and Rob attained maximum separation? No approach. A recession. Otherwise put, a negative approach velocity of -0.8×10^{-10}. The total change in approach velocity—a change from negative to positive—is twice as big as either number alone. Our measure of the bending of the two worldlines toward each other in those 2 seconds is thus 1.6×10^{-10}.

The spacetime area responsible for that bending is bounded by the timelines between Rob and Alix at the instant of the first and second speed measurement and the worldlines traversed between the two measurements. Near the center of the Earth, Rob and Alix were each traveling at a speed of 7.9 kilometers per second. Thus in 2 seconds their worldlines traversed close to 16,000 meters. To convert the timelines into the geometric units of spacetime, we simply multiply by the speed of light—(2 seconds) $\times$ (3×10^8 meters per second)—to get 6×10^8 meters. The product of the breadth in space and length in time gives the embraced spacetime area, 9.6×10^{12} square meters.

Now at last we can figure the curvature, the responsibility of each square meter of the embraced spacetime for that bending. It is the quotient of the 1.6×10^{-10} bending divided by the 9.6×10^{12} square meters of spacetime area responsible for the bending. That quotient, that per-unit-area responsibility, that curvature is 17×10^{-24} of bending per square meter of spacetime.

How can so small a curvature add up to so big an effect as the interweaving of the worldlines of Alix and Rob as they boomerang through the Earth? How can it bend their worldlines toward each other strongly enough to make the two free-float earthships repeat their minuet every 84 minutes? The answer lies not in the smallness of the curvature but in the slowness of earthlings! We, our cars, even the boomerangers' earthships creep along at the pace of a snail compared with light. The time for a round trip through the Taralga-Atlantis shaft is ungodly long according to any geometric light-travel measure of time. In compensation, we humans have cleverly adapted our language so that we won't have to see or speak the truth. A deed we call quick when we've done it in a second. But a second for light means a distance covered of 300,000 kilometers, seven trips around the Earth. And the boomerang round-trip time, 84 minutes, is more enormous yet. Oh, how slow it looks to light! Not zooming. Not walking. Not creeping. Oozing! That long slow ooze racks up an enormous number of meters of light-travel time. That number is so huge that—by the end of one step of the minuet—it can and does compensate for the smallness of the spacetime curvature.

Other Slants on Curvature

Like a great new sculpture, so curvature preserves its unity, strength, and beauty as we detect and measure it from three quite different standpoints, called *bending, transport,* and *distance.* This unity of different views of curvature we see most vividly in everyday examples of curved surfaces. Here the geometry has only two dimensions, not four. And here the medium is not spacetime, but space. It matters little whether the background for our thinking is a saddle or a boulder, a planet or a potato, a Zeppelin or a ball. So for definiteness—and simplicity—let's consider the two-dimensional surface of a ball.

Bending. We construct two nearly parallel and nearly like-directed geodesic line segments of equal length on the surface of the ball. We measure the angle of approach between the geodesics at the beginning of the segments and again at the end of the segments. The difference between the two angles of approach measures the bending—relative to each other—of the two geodesics. We express these angles and their difference not in the Babylonian unit of degrees but rather in radians.

Of the two angles of approach, one will serve as illustration: the angle of approach at the beginning. Looking right at the start of the two segments, we mark off on the one a very short bit of line and likewise a bit on the other, of equal length, both so short that each is essentially straight. First we measure the distance between the starting points of those two bits and then the distance between their end points. Next we take the difference, and finally we divide this difference by the common length of the two bits of line. The quotient directly gives the angle of approach of the two geodesics at the start of the chosen segment of their runs. Likewise we find the angle of approach of the two segments at the end of their runs. The difference in angle of approach at the two ends is the bending.

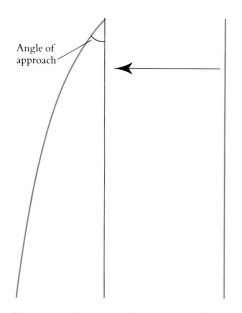

Angle of approach

Determining the angle of approach in radians between short bits of nearby geodesics. The angle of approach is the angle that would be formed if the two geodesics were brought together so that their ends touched.

RADIAN AS MEASURE OF ANGLE AND OF BENDING

To describe an ever changing direction, no measure of angle is more convenient, more useful, and more generally used than the radian. One radian is the angle subtended at the center of a circle by an arc equal to the radius of that circle.

Consider an equilateral triangle. The angle at the lower left-hand vertex is 60 degrees, any onlooker will at once see. We, however, ask what a circle looks like that has for center this lower left-hand vertex and sweeps in a great arc through the other two vertices. What is the arc length between the two sides of the triangle? One sixth of the entire circumference, one sixth of 2π times the radius, that is, $\pi/3$ or 1.0472 times the radius. By definition, that arc, divided by the radius, *is* the angle: 1.0472 radians.

Were we tempted to take a shortcut, we might use, instead of proper arc length, the straight-line distance between the two vertices not adjacent to the point where we want to know the angle. In the present case we know that intervertex distance to be equal to the radius itself. Taking it for a rough and ready estimate of the arc length, and dividing it—as definition requires—by the radius, we arrive at an approximate figure for the angle: 1.0000 radian. It's evidently low, but off only in the ratio 1.0000 to 1.0472. When we divide the one arc into two equal arcs, and approximate each of them by a straight-line run, the two straight lines add to a length of 2×0.5176 radii. Using this new and better approximation, we figure an angle of 1.0353 radians compared to the true 1.0472 radians. That subdivision from one straight run to two has cut the error from a bit more than 4 percent to a bit more than one percent. A second subdivision approximates the arc by four straight runs. It reduces the error to less than $\frac{3}{10}$ of one percent. In brief, the error, even the percentage error, incurred when we replace arc by straight line goes rapidly to zero when we deal with ever smaller angles. And small, indeed, *is* the angle by which one geodesic bends, in a short portion of its course, relative to a nearby geodesic!

Top: *One radian is the angle subtended at the center of a circle by an arc equal to the radius of that circle.* Bottom: *The vertex of an equilateral triangle positioned at the center of a circle subtends an arc of 1.0472 radians.*

We now divide our value for bending by the area responsible for it—the area between the two line segments—to get the curvature:

$$\text{space curvature} = \frac{\text{change in angle of approach of two geodesics}}{\text{space area responsible for the change}}$$

Thus we sense curvature of space by observing the bending of two geodesics toward each other. Little does it differ from finding the curvature of spacetime by measuring the change in approach velocity of the worldlines of Alix and Rob.

Transport. In imagination let a traveler shrink so much in size, and increase so much in sticking power that he can travel the surface of a ball

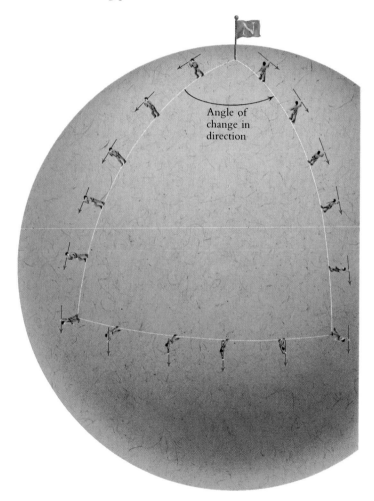

Angle of change in direction

Parallel transport of a marker around a closed circuit on a curved surface produces an angular change in direction of the marker even though the marker is never turned at all during transport.

as comfortably as can any ant. Starting at the North Pole of the ball, he travels straight south to the equator. On the way, he carries a javelin, pointed always straight ahead. At the equator, the traveler makes a 90-degree left turn but continues to point the javelin, as it was, straight south. He now marches straight east for any distance he wishes, all the way unwaveringly holding the javelin, without change of direction, so it continues to point straight south. Now another left turn of 90 degrees for the traveler, none for the javelin. The traveler is now going straight north, straight back to the starting point. Throughout this last leg of the journey the javelin continues to point straight to the rear, straight south.

Not at a single step of the way has the javelin turned one iota of a radian from the direction it had an instant before. No wonder that the transport of a direction marker with such care demands and receives a standard name: *parallel transport*. Even though the javelin remains pointed in the same direction, south, every step of the way, yet, at the completion of its parallel-transport round trip, it is turned relative to its start-off direction. Turned through how great an angle? In other words, what is the angle—measured at the North Pole—between initial and final direction of the javelin? It is identical to the angle covered at the equator by the traveler.

Do we pin all the blame for this change in direction on the traveler? No. He could not have done his job more conscientiously. Moreover, had he carried the javelin in parallel transport around a closed circuit on a flat surface, a plane, a two-dimensional geometry free of all curvature, then that pointer would have returned to its starting point with no change in direction at all.

Then do we pin all the blame for the actual change in the pointing of the javelin on the curvature of the ball? No. That curvature would have been without effect on the pointing of the direction indicator if the traveler had never made his trip, never explored the surface of the ball, never circumnavigated all those square centimeters of its landscape.

Where, then, must we fix the responsibility for the net change in direction of a pointer transported parallel to itself around a closed circuit? Jointly on the curvature of the surface and the amount of surface circumnavigated by the circuit. The shape of the circuit doesn't matter: it can have the most complicated shape, provided only that it is closed.

To determine the curvature, we divide the change in direction of the javelin—carried in parallel transport around the perimeter of a local region—by the amount of area circumnavigated:

$$\text{space curvature} = \frac{\text{change in direction during parallel transport}}{\text{space area circumnavigated}}$$

Thus to consider change of direction in parallel transport is to have a second means to define and measure curvature. A third means comes from looking at distance, which we glanced at in Chapter 1.

Distance. The idea in brief? Discrepancy in circumference reveals curvature! The circumference of a circle traced on a curved surface doesn't come out equal to what's expected for a circle on a flat surface. No example lies closer to hand than a spherical ball. Run a string around its middle. Thus find, let's say, that the circumference of the ball at the equator is 40 centimeters. The ball being spherical, we will find the same circumference when—again with the string—we measure distance from North Pole to South Pole and from there around back up to the North Pole.

Now fold the 40-centimeter string on itself twice, so that it becomes only a quarter as long as it was. Fasten one end of the resulting 10-centimeter cord at the North Pole by sticking a pin through the cord and into the ball. Through the loop at the other end of the cord, put the tip of a pencil. With that pencil draw a line all the way around the ball. Thus we can trace a circle on the ball's surface—not any circle, but the circle that we call the equator. As we draw that curve, we rack up a length of 40 centimeters. Now take the same 10-centimeter cord and pencil and draw a circle on a flat sheet of paper. We get a circumference given by the familiar expression ($2\pi \times$ radius)—in our example, $2\pi \times 10 = 62.8$ centimeters. The curvature—the positive curvature—of the ball's surface shortens the circumference by 22.8 centimeters. That 22.8-centimeter shortening is the discrepancy in circumference between the two circles.

What is the origin of this discrepancy? In brief, this. In order to fit the flat sheet of paper over the curved surface of the ball we have to scissor away "extra" unwanted spearlike segments of paper, thus separating one "orange-peel sector" from the next. When we throw away those portions of the sheet of paper, we also eliminate a sizable fraction of the original circumference. We end up with a decorative pattern of orange-peel segments. Each connects only at its vertex to its two nearest neighbors. As we smooth the paper pattern down over the surface of the ball, the portions of the old circumference that still remain all pull in toward each other. They join up to make a continuous curve contiguous with the ball's equator. That is one way to see why the equator of a sphere has a circumference that is shorter than it would be if the surface were flat.

The negative curvature of a saddle reveals itself in the *lengthening* of the circumference of a circle relative to the flat-space expectation. Why a lengthening? A look at the drawing on the next page reveals the answer. At the portions of the saddle closest and farthest from us lie the two regions where the surface curves down. To our left and right lie the two areas

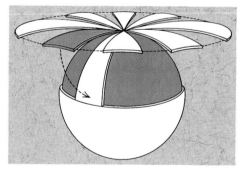

A flat circle of radius r *cannot be made to fit smoothly over the curved surface of a ball unless "extra" wedged-shaped pieces are removed. The total length of the flat circle's circumference so removed equals the shortening in the circumference of a circle traced with a geodesic of length* r *on the ball's surface.*

Chapter Five

where the saddle curves up. We take the saddle point itself as the center and, with our cord and pencil, draw a circle with a 10-centimeter radius on the saddle's surface itself. This maneuver requires ingenuity to keep the taut radius-measuring cord from lifting off the surface in those places where the saddle curves up. Were that circle drawn on a flat surface, it would have a circumference of 62.8 centimeters. Compared with that flat-space circle, however, the saddle's circle—like a roller coaster—rises and falls twice as it goes once around its course. So its actual circumference exceeds the flat-space expectation. This lengthening of the circumference tells us that the saddle has a negative curvature.

We can do more with the method of circumference discrepancy than merely distinguish between positive (spherelike), zero (flat), and negative (saddlelike) curvature. We can determine the magnitude of the curvature.

First we divide the circumference discrepancy by the flat-space figure for the circumference to get the fractional circumference discrepancy. Then we divide this fraction by the circle area responsible for it. Finally we multiply the resulting quotient by 6π in order to make the numerical factor come out right. It's simple enough to carry through these two divisions and this one multiplication. It's natural, too, to take the circle in the first place to be small enough so that we're dealing with a reasonably *local* region of the surface. This granted, we end up with a good figure for the curvature:

$$\text{space curvature} = \frac{\text{fractional circumference discrepancy}}{\text{space area responsible for discrepancy}} \times 6\pi$$

All three methods of determining the curvature of a two-dimensional space geometry give values in good agreement for a given surface.

As an example, take the ball with its 40-centimeter circumference. The bending, transport, and distance methods all give for the curvature of its surface the same result, 0.0247 radian per square centimeter, or 247 radians per square meter.

The average curvature, by contrast, of the two-dimensional space that marks the surface of the Earth is only 2.5×10^{-14} radians per square meter. Two travelers who start off from nearby points on the equator and travel straight north have to embrace between their two geodesics many square meters of landscape before they notice much bending of those geodesics toward each other.

The curvature of the two-dimensional *spacetime* in which Alix and Rob boomeranged is yet smaller, only 1.7×10^{-23} radians per square meter—smaller by nine orders of magnitude. No wonder that it takes many meters of time—42 minutes is 7.6×10^{11} meters of light-travel

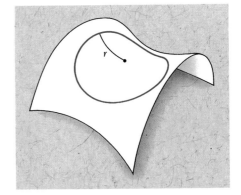

The circumference of a circle traced with a geodesic of length r on a saddle-shaped surface is longer than the circumference of a flat circle with the same radius, r.

time—to bend their worldlines, their geodesics, their tracks through spacetime, from goodbye to hello!

Spacetime Curvature Inside and Outside the Earth

How does spacetime curvature outside the Earth—or any other concentration of mass—differ from spacetime curvature inside? That difference between outside and inside—how do we bring it into evidence? And what light does it shed on Einstein's great idea that gravity is a local-only manifestation of spacetime curvature?

The curvature revealed by the bending of the geodesics of Alix and Rob toward each other within the Atlantis-Taralga boomerang shaft is the curvature of a two-dimensional spacetime, built on the time dimension and a single in-and-out space dimension. Now we reach for a broader perspective, and on the way to it uncover four new aspects of curvature: (1) inside the Earth a spacetime curvature that is *contractile;* (2) outside the Earth a spacetime curvature that is *tide-driving;* (3) a *tear-free* response of spacetime geometry in any region to bending of the geometry in the regions immediately adjacent to it; and—in consequence—(4) a *dilution* of response to the bending effect of mass in regions ever more remote from that mass.

How are we to see these new features of curvature? We must break out of the narrow view of a one-dimensional channel through the Earth. Out of a spacetime of only two dimensions: one of them past-future, the other in-and-out along the shaft. We must somehow sense spacetime in its full four-dimensionality: east-west, north-south, in-out, yesterday-tomorrow. *That* is the real spacetime around us, through which, minute by minute, all our separate worldlines thrust their way ever further forward. Creating in the mind a single vivid image of four-dimensional spacetime is beyond the picture-making power of anybody I know. But is it really any easier to capture from one view all the richness of a modern metropolis? That power no single sight possesses. We carry in the memory instead a hundred images. Each registers on the retina from a separate vantage point. In a similar way we can grasp spacetime curvature. And of curvature in a given locale, we need not a hundred but only six views. Those views—each of them a two-dimensional slice of the full four-dimensional spacetime—I call *plaquettes*.

Why are there six views, six natural choices of plaquette? Because, from a spacetime of four dimensions, there are six ways to select out two of those four dimensions for special attention. Our first example, the boomeranger's, gives us a plaquette in which time and in-out direction are the two dimensions. A second example, leading to another view of the curvature of the spacetime in and around Rob's capsule? Select another two of the four dimensions, another plaquette—say, time and the east-west direction. To explore the curvature of this second plaquette, recall how we conducted the exploration of the previous spacetime, time–plus–in-out dimension. We capitalized on the presence of Alix zooming through the center of the Earth, 16 kilometers behind Rob, along the in-and-out Taralga-Atlantis shaft. With her help we could get radar measurements of the slowly changing Rob-Alix distance, and so determine the bending of those two geodesics toward each other.

Now we need a third boomeranger, Fido, 16 kilometers off to the east, that is, off to the left of Rob. Fat chance! There's no shaft there! Spend more money? Dig it? No. Too expensive! The direct opposite. Give up all this costly shaft business, despite its romance. Replace Rob, Alix, Fido, and all the rest of our doughty explorers by test masses so small and so penetrating that they don't require for their free passage the digging of any shafts at all. Where can we beg, borrow, buy, or steal those test masses? I can't tell you where; so imagine them. But call them by the old names!

Rob and Fido's 16-kilometer sideways separation at the center of the Earth, like Rob and Alix's 16-kilometer in-line separation, is temporary. It begins to decrease, at first slowly, then more rapidly. In other words, the Rob-Fido pair of free-float worldlines also bend toward each other. This bending reveals that the curvature of the time–plus–east-west two-dimensional space is positive. Moreover, this curvature—inside an ideal Earth of effectively uniform mass density—is identical in magnitude to the curvature in our earlier time–plus–in-out two-dimensional slice through spacetime.

Let's take a third two-dimensional slice through spacetime in the vicinity of Rob, a slice specified by time plus distance between Rob and yet another test mass, Pep. This test mass at the start of observations is 16 kilometers to the north of where Rob is. When we follow their slowly gathering speed of approach, we discover the same positivity of curvature and the same magnitude of curvature as in the other two slices through spacetime. Summary: test masses near the center of the Earth, originally somewhat separated from each other in whatever direction, and originally neither approaching nor receding from each other, gradually begin to approach.

We come thus to our first new finding about curvature. Spacetime curvature inside the Earth is *contractile*. It bends nonapproaching free-float worldlines into approaching worldlines. In stronger language, mass inside the Earth bends the spacetime inside the Earth, forcing on it curvature of a contractile character. Moreover, the strength of this curvature is

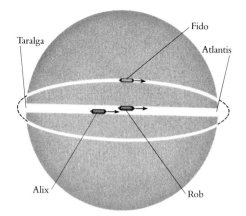

A third test mass, Fido, zooms through the Earth to the east of Rob.

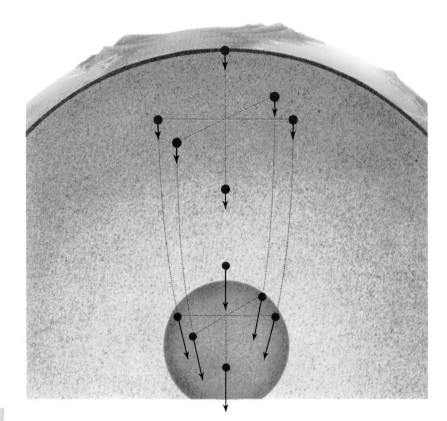

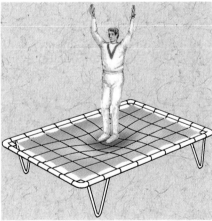

The deformation of the trampoline fabric by the impact of a jumper is analogous to the bending of spacetime geometry inside the Earth by the Earth's mass.

the same just below the surface of an ideal Earth of uniform density as it is at its center. Good!

To visualize this contractile feature of curvature, it's fun to compare spacetime geometry with the fabric of a trampoline. The jumper bends the fabric of the trampoline in the zone of impact; likewise Earth's mass bends spacetime geometry inside the Earth. The fabric in the region of the impact sustains a cuplike deformation; so too the spacetime geometry within the Earth suffers a contractile bending.

Beyond thus making plausible our first new finding about curvature—contractility—the analogy between spacetime and a trampoline provides us with a happy doorway to a second new insight. The deformation of the trampoline in the region where the jumper lands brings about a curving of the fabric even outside the zone of impact. Likewise the bending of spacetime geometry *inside* the Earth forces bending on the spacetime geometry *outside* the Earth. For that reason spacetime finds itself curved in regions totally free of mass.

Chapter Five

The character, however, of spacetime curvature outside the Earth is different from what it is inside. The trampoline fabric, too—blessed guide to our thinking—has a curvature outside the zone of impact different in character from what it is inside. Outside—saddlelike and therefore negative. Inside—spherelike and therefore positive. The difference in spacetime curvature inside and outside the Earth is equally striking. Inside, contractile. Outside, noncontractile. Noncontractile but *tide-driving!* That is new insight number two on curvature.

How can we see this tide-driving character of spacetime curvature outside the Earth? Set into free float a cluster of widely spaced and tiny test masses. At release, dispose them in the configuration of a hollow sphere, a sphere 32 kilometers in diameter. This free-float sphere, this array of tiny test masses at first changes neither in size nor in shape. Now watch! See the sphere begin to change into an egg. Two dimensions shrink. The third lengthens, and twice as fast. The consequence? Zero change in the volume of the cluster. In this sense, a totally noncontractile spacetime curvature. Spacetime curvature of a character that's pure tide-driving—it drives a tide, in the sense that it stretches out what was a sphere into an egg. The tide rises because spacetime curvature caused by Moon and Sun has pulled the layer of ocean covering much of the Earth into an egglike pattern that rises higher at the ends than in the middle.

The tide-driving spacetime curvature outside the Earth owes its existence to and is forced by the contractile bending of spacetime within the

An array of test masses covering the surface of a hollow sphere freely floating above the Earth's surface will shrink in two dimensions and lengthen in one; the volume remains constant, only the shape changes. This change is evidence of the noncontractile, tide-driving spacetime curvature outside the earth.

Earth. If spacetime did not thus respond, it would tear! Where can we see evidence of this *tear-free* fit of geometry inside to geometry outside? Look at two free-float test masses whose line of separation runs parallel, or tangential, to the Earth's surface, test masses that float downward at the same up-down level, masses that fall from just above the Earth's surface to just below. Their geodesics, their free-float world lines, their tracks through spacetime bend toward each other equally strongly outside and inside. In other words, spacetime curvature—in the two-dimensional spacetime defined by time and tangential direction of separation of those two masses—is positive in sign and of equal strength just inside and just outside the Earth. Spacetime geometry inside and outside fit together without a tear.

Loosely stated, curvature tangent to the surface fits, but curvature perpendicular to the surface doesn't. What does curvature "perpendicular to the surface" mean? That's shorthand for curvature in the two-dimensional spacetime defined by time and by the direction in-and-out from the center of the Earth. Throughout the interior of an ideal Earth of effectively uniform density, this perpendicular curvature is identical in strength with curvature in other directions, is positive, and is independent of position. We saw this equality earlier when we compared the worldlines of two test masses, or boomerangers, oriented in different directions (in-out, east-west, or north-south) inside the Earth. Outside the Earth, however, the perpendicular curvature is negative and twice as strong as the curvature in the east-west direction, the north-south direction, or any other direction tangent to the surface. In other words, counting all three directions together, spacetime curvature outside the Earth has zero net volume-shrinking power and so rates as completely noncontractile.

In brief, where there's mass, there curvature is contractile. Where there's no mass, there curvature is noncontractile. Contractility of spacetime curvature, here, responds to mass density, here. Contractility there is proportional to mass density there. Whatever the place, the density at that place governs the contractility of spacetime geometry at that place. This feature of nature sends us a message about the grip of mass on spacetime. We already begin to see in broad outline how gravity works.

Mass *here* makes its influence felt *there* by a spreading network of influence. Every link in this network is spacetime geometry. The spacetime inside the Earth, impressed with contractile curvature by the mass inside the Earth and forbidden to tear, communicates a like *tangential* curvature to the immediately surrounding region of spacetime. Spacetime there, free of mass, is free of contractility. The curvature that spacetime there acquires in the two tangential directions has to be balanced out by curvature that is twice as strong in the third perpendicular direction—that is, curvature in the time–plus–in-and-out two-dimensional geometry.

The shell of emptiness outside the Earth in like manner communicates curvature in the two tangential directions to the next-farther-out empty spherical shell of spacetime. In other words, spacetime responds by yielding in region after successively more remote region. The nature of this response? Tide-driving curvature, not contractile curvature.

The tide-driving curvature of spacetime grows ever weaker the more distant from the central mass. Why? It grows weaker by *dilution*. The greater the distance from the mass, the more enormous is the region over which the central bending influence is felt!

Twice the distance from the center of the mass? Then the volume over which the influence is felt is doubled in extent in all three space dimensions; the volume is enlarged by a factor of 8 ($2 \times 2 \times 2 = 2^3 = 8$). At that distance, spacetime bending will be diluted to one-eighth of what

THE GEOMETRIC MEASURE OF MASS

Mass submits to being measured, interpreted, and understood as length—and as surely as does time. Only after 1905 and Einstein's special relativity did mankind fully appreciate that a time is a length, that the phrase "second of time" is an antique way to speak of 3×10^8 meters of light-travel time. Only after 1915 and Einstein's geometric account of gravity did mankind fully appreciate that a mass also is a length, that the phrase "kilogram of mass" is an antique way to speak of 0.74×10^{-27} meters of mass. The mass of the Sun in conventional units is 1.99×10^{30} kilograms. That's 1476 meters in the geometric measure of mass that Einstein taught us to use. That's about a mile. It's also—so it happens—half of the reduced circumference that the Sun would have if that star were collapsed into a black hole—that is, a mass so compact that light cannot escape from it. The mass of the Earth, 6.0×10^{24} kilograms in conventional units, is 4.44×10^{-3} meters in geometric language. That's 4.44 millimeters, about the diameter of a pencil lead. With so small a mass and so great a radius (6.37×10^6 meters), it's no wonder that an ideal Earth of uniform density produces throughout its interior—and in the Taralga-Atlantis boomerang shaft—a contractile spacetime curvature that's so small: mass/(radius)^3 = 1.7×10^{-23} bending per square meter of spacetime embraced. Because the Earth's mass is so "dilute," and the resulting spacetime curvature so small, it is no wonder Rob and Alix take a long time by the standards of light (84 minutes = 1.5×10^{12} meters of light-travel time) to complete one cycle of their minuet.

it is at the Earth's surface. More generally, and in keeping with this reasoning, the curvature falls off inversely as the cube of the distance. Still, even 400,000 kilometers away from the center of the Earth, the tide-driving effect of the mass of the Earth is strong enough to deform the shape of the Moon out of sphericity by about a kilometer! The Moon, by a like influence, produces at the distance of the Earth a tide-driving effect well-known to every watcher of the ocean.

It's clear enough what we mean by distance from the center of the Earth. But what do we mean by distance from the center of a black hole, where every attempt at a measurement is bound to fail? An answer that's standard for any object, whether Earth, neutron star, or black hole, that has an essentially spherical symmetry: tote up the circumference of a circle, all places on which display the same curvature, and divide this circumference by $2\pi = 6.28$. Call the resulting distance figure the reduced circumference or, equivalently, the Schwarzschild r-coordinate.

With the help of the breath-saving, universally used abbreviations r for distance and m for mass, the tide-driving curvature of spacetime *outside* the Earth shows itself as

$$\begin{pmatrix} \text{curvature in plaquette} \\ \text{defined by time and} \\ \text{in-out direction} \end{pmatrix} = \frac{-2m}{r^3}$$

$$\begin{pmatrix} \text{curvature in plaquette} \\ \text{defined by time and} \\ \text{by any direction} \\ \text{perpendicular to} \\ \text{in-out direction} \end{pmatrix} = \frac{m}{r^3}$$

$$\begin{pmatrix} \text{total} \\ \text{contractile} \\ \text{curvature} \end{pmatrix} = \begin{pmatrix} \text{curvature in} \\ \text{in-out plaquette} \end{pmatrix} + 2\begin{pmatrix} \text{curvature in} \\ \text{perpendicular} \\ \text{plaquette} \end{pmatrix} = 0$$

The contractile curvature of spacetime *inside* the Earth is

$$\begin{pmatrix} \text{curvature in plaquette} \\ \text{of any orientation} \\ \text{anywhere } inside \text{ ideal} \\ \text{Earth of uniform density} \end{pmatrix} = \frac{\text{mass}}{(\text{radius of Earth})^3}$$

The Earth's mass, right where it is, grips spacetime, right there, and impresses on it a contractile curvature. Thus bent inside the Earth—yet tear resistant—spacetime outside also has to bend. The bending there, however, being in a region totally free of mass, is totally free of contractility. The *local contractility* of curvature, then, signals, marks, and measures the *local grip* of mass on spacetime! Can we find any equally clear indicator of the grip of spacetime on mass? Yes, in what happens when two masses smash into each other, a topic we will explore in Chapter 6.

The Tide-driving Power of Moon and Sun

The ocean's rise and fall in a never-ending rhythmic cycle bears witness to the tide-driving power of the Moon and Sun. In principle, those influences are no different from the spacetime curvature outside the Earth. When that curvature acts on a thin spattering of free-float test masses, spherical in pattern, it draws them out into an egg. By like action the Moon, acting on the waters of the Earth—floating free in space—would draw them out into an egg-shaped pattern if there were water everywhere, water of uniform depth. There isn't. The narrow Straits of Gibraltar almost cut off the Mediterranean from the open ocean, and almost kill all tides in it. Therefore it is no wonder that Galileo Galilei, although a great pioneer in the study of gravity, did not take the tides as seriously as the more widely traveled Johannes Kepler, an expert on the motion of the Moon and the planets. Of Kepler Galileo even said, "More than other people he was a person of independent genius . . . [but he] later pricked up his ears and became interested in the action of the moon on the water, and in other occult phenomena, and similar childishness."

Foolishness indeed, it must have seemed, to assign to the tiny tides of the Mediterranean an explanation so cosmic as the Moon. But mariners in northern waters face destruction unless they track the tides. For good reason they remember that the Moon reaches its summit an average 50.47 minutes later each day. Their own bitter experience tells them that, of the two high tides a day—*two* because there are two projections on an egg—each also comes about 50 minutes later than it did the day before.

Geography makes Mediterranean tides minuscule. Geography also makes tides in the Gulf of Maine and Bay of Fundy the highest in the world. How come? Resonance! To understand why, let's visit Kip and Carolee as they lunch in a coastal Nova Scotian restaurant.

The waitress approaches their table pacing in time with a cheerful march. On her head she carries a tray laden with a huge bowl of chowder for Carolee, a lesser bowl of soup for Kip, and two cups of coffee. "So sorry," she apologizes to Kip as she wipes the soup off the plate under his bowl. "You people would call that resonance, wouldn't you? I, though, call it tuning because the beat of that march and my footsteps, friend, were in tune with the slosh of the soup in your bowl. No wonder that's what slopped over, not the chowder, not the coffee. Right?"

No greater tides are there to be seen anywhere than those at the Hantsport Pier of the Minas Basin Pulp and Power Company in Nova Scotia at the head of the Bay of Fundy. Between 8:30 A.M. (left) and 1:30 P.M. (right) on January 29, 1979, the tide rose 16 meters. Four factors contributed to this unusually high tide: (1) The Moon was at its closest perigee of the year on January 28, (2) the Moon was "new" on January 28, (3) Earth was near perihelion, and (4) the barometric pressure (about 97 kilopascals) was low.

Kip smiles, "Right—resonance! And resonance, Miss, explains why we're here in Nova Scotia, in this town of Hantsport, near the pier of the Minas Basin Pulp and Power Company. We are here because it's the best place in all the world to see the tides at their highest."

"And how come resonance makes our Bay of Fundy tides so high?" she asks.

"Because the natural time of slosh of these several hundred kilometers of Gulf of Maine water hereabouts is close to 13 hours. That's in almost perfect tune with the 12.4-hour timing of the Moon's tide-driving power—even with the 12-hour timing of the Sun's influence. Resonance indeed!"

Carolee breaks in to put the idea in other words. "A bay is to an ocean as a bowl is to a luncheon tray. The to-and-fro of the tray as you marched toward our table drove the liquid in the containers. The up and down of the tide in the great broad ocean likewise drives the waters in every bay that gives upon the sea. A small bay has a quick slosh time. So there is no resonance with the 12-hour 25-minute period of the tides. But in the larger Bay of Fundy a longer interval is needed for the water piled up at one end to rush to the other and back. That slosh time is marvelously tuned to resonate with the far smaller rise and fall of the open ocean."

As the waitress moves on to serve those at the next table, Carolee and Kip begin eating their lunch and discussing their visit to Nova Scotia. Because these tide-tracking travelers wanted to see the very highest tides, they had to pick carefully the best time for their trip. They had to come in summer, the time of year when the Bay of Fundy, a northern body of water, tilts most strongly toward the Moon, and when the Earth comes closest to the Sun. They also had to remember that the Moon in its elliptic orbit comes roughly 10 percent closer to the Earth at some times than at others, and has roughly 35 percent greater tide-producing power at its closest than it does at its most remote. Finally, they had to take into account the sun's tide-producing power, which is about 45 percent as great as the tide-producing influence of the Moon. The Sun's effect reinforces the Moon's influence when the Moon is full—and also when the Moon is dark, dark because interposed, or almost interposed, between Earth and Sun. At this time, the total tide-driving influence acting on the Earth from outside is about 1.45 times as great as the Moon alone would give.

Having picked a time that would assure the highest tide, Kip and Carolee also were assured of seeing the lowest tide. Indeed, when they had visited the pier that morning, about 8 o'clock, all the dorys, fishing boats, and ships were lying over on their sides with no water in sight to keep them afloat. Now, as Kip and Carolee finished their lunch, it was time to head back to the pier for the highest of high tides, at about 2:30 P.M. When they arrived, the water had risen 16 meters higher than where it stood in the morning. The vast reach of that newly arrived water, repeated roughly twice each day, gives spectacular evidence of the tide-driving power of Sun and Moon, a power that originates in the grip of faraway Sun and Moon on faraway spacetime—and the grip of the thus influenced spacetime here on water here.

6 Momenergy: The Grip of Spacetime on Mass

Oh, mass in motion,
On spacetime's great stage,
Fields of force of every kind,
You show us clear and bright
By the rule of motion that you follow
The handle by which spacetime grips you —
Momentum-and-energy,
Momenergy.

You, mass over there,
Like any other mass,
You reveal to us your momenergy
By an arrow located in spacetime
Right where you are at this very instant,
An arrow of length equal to your mass.

You, two masses nearby,
Crashing into each other
Change, each of you, your individual momenergy.
But the magic grip of spacetime on momenergy
Demands, ensures, and guarantees
That the sum of your two arrows of momenergy
Stays ever constant.

You, billions of masses in wild encounter
Within yonder great explosion,
Spacetime's grip on you ensures
That never in any local 3-volume
Followed for any little time,
Is there ever any creation of momenergy.

Four trades of momentum and energy between fast-moving protons (red) and identical hydrogen nuclei, originally at rest, in a liquid hydrogen bubble chamber.

The bigger the mass, the bigger its voting power. Or so at least justice would seem to demand. Shouldn't two like masses, bundled together, exert twice as much grip on spacetime as one does? Ha! So they do! This we discovered in the last chapter. There we found that spacetime curvature is proportional to the mass that produces that curvature. Conclusion: the voting power of mass shows up in a reasonable way when we look at the grip of mass on spacetime.

When we turn to the other side of the story of gravity, however—the grip of spacetime on mass—we seem at first sight to find mass stripped of any voting power whatsoever. To mass, spacetime says, "I am the geometry in which you live and move. Curved though I am, I offer you the choice of this, that, or the other geodesic. Pick one! Having picked it, stay on it! Thus guarantee to yourself, oh mass, a free-float existence, a worldline that will carry you from the past through the present and on into the future with never a jar or jolt." Not one word here about the magnitude of the mass that's going to live out its life on that worldline. Not one iota of distinction do these instructions make between the motion of a big mass and the motion of a little one.

Is mass in truth totally subject to the governance of spacetime? Is mass really deprived of all voting power? No! For evidence, look at the crash of two masses, two spaceships, zooming through the emptiness of space. Up to the instant of the crash, each pursues its chosen worldline. The trueness of their courses seems to show that each acknowledges no other sovereign than spacetime. In the crash, however, each changes course. Spacetime alone no longer guides each in its motion. A force is at work between the two, a force that throws each from its course. In this yielding by spacetime of its full sovereignty, we discover what it still

Spacetime grips mass, keeping an object moving straight when free. By its power, it enforces conservation of energy and momentum in a smash. The coupling of mass and geometry, far from being the weakest force in nature, is the strongest.

Chapter Six

insists on and what it concedes. Most important, we discover that space-time grips each mass with a strength proportional to the magnitude of that mass.

The Conservation of Momenergy

Paradoxically, few features of motion are more complicated than a collision, and few are simpler. The complication shows nowhere more clearly than on the slow-motion videotape of the smashup of two spaceships. Millisecond by millisecond the shell of one colliding spaceship is pushed in another centimeter. Millisecond by millisecond the fuselage of the other spaceship bends in a little more on the way to total collapse. Steel against steel, force against force, crumpling surface against crumpling surface. What could be more complex than the detailed analysis of all that metal wrinkling?

For the pilots of the colliding spaceships the experience is shattering. They are hardly aware of the noise and the complicated damage. A single impression overpowers their senses: the inevitability of the crash. Call it what we will—inertia, momentum, the grip of spacetime on mass—something is at work that drives the two spaceships together as the frantic astronauts crash into one another and then rebound.

Does spacetime cease its grip on mass during the collision? No. It does its best to keep each spaceship going as it was, to keep its *momentum* constant in magnitude and direction. The momentum of an object we can think of loosely as that object's will to hold its course, to resist deflection from its appointed way. It measures the object's power to impart a kick in a collision. The higher its momentum, the more violently it hits whatever stands in its way. But although spacetime tries to keep the momentum of each spaceship constant, its grip is not all-powerful. Inertia cannot prevent the two vehicles from exchanging some momentum. It does demand, however, that whatever momentum one spaceship gains, the other spaceship has to lose. Regardless of all complications of detail and regardless of how much the momentum of any object individually may change on impact, the combined momentum of the separating objects is identical to the combined momentum of the objects that collide. A like statement applies to *energy*, despite a conversion of energy of motion into heat energy and fuselage-crumpling.

A collision thus manifests a wonderful simplicity: the combination of the motion-descriptive quantities (momentum and energy) of the two colliding bodies does not change. That combination is identical in its value before and after the collision. In a word, it is "conserved." This conserved combination I call *momentum-energy* or, more briefly, *momenergy*. I will use these two terms, momentum-energy and momenergy, interchangeably in the remainder of this book.

It is one of the great miracles of nature that spacetime grips mass in such a way as to conserve the momenergy of our crashing spaceships. It is impossible, however, for spacetime to grip mass without mass also gripping spacetime, affecting it, modifying its ideal flatness, curving it. This tit-for-tat grip is the golden door to gravity. However, under everyday circumstances the resultant bending of spacetime is so small, the gravitational effect of one spaceship on the other is so unbelievably weak, that we generally neglect it. Only very large masses can bend spacetime appreciably.

The miraculous action of spacetime on mass, however, we neglect at our peril. The pilot of each spaceship, unrestrained by any seat belt, pitches into the cabin wall as one vehicle whangs into the other. Each ten-ton vehicle itself, after the crash, continues to zoom onward through space, but with an altered velocity and in a new direction. A collision cannot be boiled down from mere talk to numbers without adopting, directly or indirectly, the principle of conservation of momentum and energy. In the enterprise of identifying the right numbers, using them, and understanding them, no concept is more powerful than what relativity smilingly holds forth: momenergy.

Words promote insight. If the word spacetime *symbolizes the unity of space and time better than the* space-time *of the compulsive hyphenator, then the word* momenergy *may speak for the indissoluble natural unity of momentum and energy better than does the hyphenated* momentum-energy.

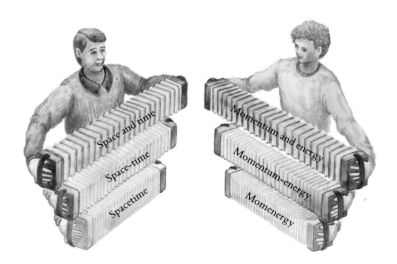

Properties of Momenergy

Momenergy is only a name. What counts are its properties. We can most easily uncover four central properties of momenergy by combining everyday observation with momenergy's essential defining feature: that its total is conserved in any collision.

First, consider two pebbles of different sizes moving with the same speed toward the windshield of a speeding car. One bounces off the windshield without anyone noticing; the other startles the occupants and leaves a scratch. The difference in their kick-delivering capacities is clear. If the second pebble has exactly five times the mass of the first, then it also has exactly five times the momenergy of the first. Momenergy, in other words, is proportional to mass. Two objects of the same mass traveling with the same speed in the same direction have identical momenergy.

Second, the momentum-energy of a mass in motion depends on its direction of travel. A pebble coming directly from the front will take a bigger chip out of the windshield than a pebble of equal mass and speed glancing off the windshield from the side. Therefore momenergy is not measurable by a mere number. It is a directed quantity. Like an arrow of a certain length, it has magnitude and direction. Our experience with the unity of spacetime leads us to expect that this arrow will have three parts, corresponding to three spatial dimensions, and a fourth part corresponding to time. In what follows we will indeed find momenergy to be a four-dimensional arrow, the *momenergy 4-vector*. Its three space parts represent the momentum of the object in the three chosen space directions; its time part represents energy.

Third, the momenergy arrow of a mass at a given event P in the course of its motion points in the same "direction in spacetime" as the worldline of that object itself (see diagram on page 98). There is no other natural direction in which it can point! Spacetime itself has no structure that indicates or favors one direction rather than another. Only the motion of the particle itself gives a preferred direction in spacetime.

A particular straw in a great barn filled with hay has a direction, an existence, and a meaning independent of any measuring method imagined by the human who stacks the hay or by the mice that live in it. Similarly, in the rich trelliswork of worldlines that course through spacetime, the arrowlike momenergy of a mass has an existence and definiteness independent of the choice—or even use—of any free-float frame of reference.

No frame of reference? Then no instruments around to measure time! Or none except the one that the particle itself carries, its own

At any given event in spacetime (P, Q, R) the momenergy of a mass in motion points in the same direction as its worldline. This is true not only in the two-dimensional spacetime depicted here but also in four-dimensional spacetime.

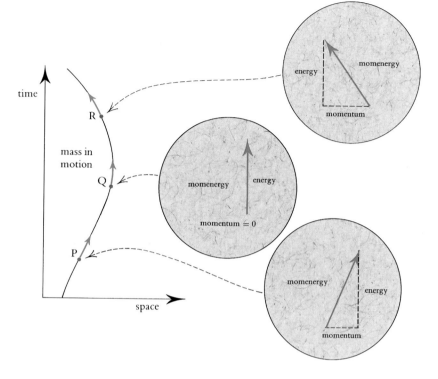

"wristwatch," its *proper time.* This discovery brings us to the fourth and final feature of momenergy. Let's imagine a particle emitting a series of flashes, P, Q, R, S, as it zooms along. If event P is followed in a very short interval of proper time by event Q, then the displacement in spacetime from P to Q is also very small. If the interval of proper time between P and R is twice that between P and Q, then so is the spacetime displacement. What governs the momenergy of the particle—besides mass—is evidently not its spacetime displacement alone, nor the amount of proper time required for that displacement to occur, but the ratio of the two. This ratio is the only directed quantity available to us to describe the *rate of motion* of the particle through spacetime. By multiplying particle mass by this proper-time rate of particle displacement, we arrive at the natural way to determine the frame-independent, or *frame-invariant arrowlike* geometric quantity that we call particle momenergy:

$$\begin{pmatrix} \text{momenergy of object} \\ \text{in free-float motion} \end{pmatrix} = \text{mass} \times \frac{\text{spacetime displacement}}{\text{proper time elapsed}}$$

No briefer, simpler, or more revealing expression has ever been discovered for momenergy, that physical quantity whose conservation governs every encounter in the known universe.

In the fraction on the right every term is familiar. "Spacetime displacement" is simply another way to speak of the "separation in spacetime between two events." And as we learned in Chapter 3, the magnitude of that separation equals the invariant quantity "interval," which is just another name for "proper time elapsed." Thus, the numerator and denominator of the right-hand term are identical in magnitude. The fraction has no purpose but to give the direction in spacetime of the arrow of momenergy; the magnitude of the spacetime displacement is irrelevant to momenergy. And if the fraction gives only the direction of the momenergy arrow, the mass must give its length. The magnitude of an object's momenergy is that object's mass.

In the previous chapters, we have seen how to express length, time, and mass in the common geometric unit of meters. By so doing, we can best emphasize the unity of spacetime. Likewise, we can best clarify the unity of momenergy by expressing all the terms in the momenergy equation in meters. When we do this, then we find momenergy itself expressed in meters.

Space and Time Components of Momenergy

Accidents of history have given us, not the one word momenergy, but two words, momentum and energy, to describe mass in motion. Before Einstein, mass and motion were described not in the unified context of spacetime but in terms of space and time separately, as that division shows itself in some chosen free-float frame of reference, some laboratory. Most of the time we still think in those separated terms. But in the same way that we unite the position of an event in space and the time of its happening in the single concept of spacetime event, so we now bring together the momentum and the energy of a moving object into the single idea of a momenergy arrow.

The unity of momenergy dissolves—in our thinking—into the separateness of momentum and energy when we choose a free-float frame of motion. From that frame let's look at the nearby flashes, P and Q, emitted by our moving particle. In that frame, the *spacetime* separation between the nearby events P and Q on the worldline of the particle resolves itself into four different separations: one in laboratory time and one in each of three perpendicular space directions. Each separation represents a separate part, a separate component, of momenergy in the laboratory free-float frame:

$$\text{energy} = \text{mass} \times \frac{\text{increment of laboratory time}}{\text{proper time elapsed}}$$

$$\left(\begin{array}{c}\text{component of momentum} \\ \text{in first space dimension}\end{array}\right) = \text{mass} \times \frac{\text{space displacement in first direction}}{\text{proper time elapsed}}$$

The component of the momentum in the remaining two perpendicular space directions is given by similar expressions.

We can picture the momenergy of a particle, a spaceship, or a planet most vividly as a spacetime arrow. This momenergy arrow has an existence and direction—in that great haystack of worldlines and events that

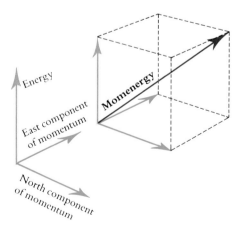

The arrow of momenergy of a moving object in spacetime can be broken down into an energy component (corresponding to time) and three momentum components (corresponding to three spatial dimensions of which one, the up component, happens to be zero and is therefore omitted from the diagram). Like the timelike interval separating the positions of two events in spacetime, the length of the momenergy arrow is invariant, that is, independent of the reference frame. This length measures the object's mass.

Chapter Six

we call spacetime—independent of the choice, or even presence, of any free-float reference frame. In contrast, the separate momentum and energy components *do* depend on the choice of frame. Nevertheless, in whatever frame seen, this arrow is timelike. In other words, momenergy has some resemblance in character to the timelike interval between two events in spacetime (see page 37), and its magnitude can be found similarly. *The magnitude of the arrow of momenergy of any object is the mass of that object.* The square of the mass is still more readily described in terms of energy and momentum—rather, in terms of their squares, thus,

$$(\text{mass})^2 = \left(\begin{array}{c}\text{magnitude of}\\\text{momenergy arrow}\end{array}\right)^2 = (\text{energy})^2 - (\text{momentum})^2$$

The momentum of a car or spaceship depends on the frame in which we see it. In one frame, terrifying. In another frame, tame. In a co-moving frame, zero momentum. In other words, in a frame moving along with the object, all the space parts—the momentum components—of momentum-energy shrink to extinction, as the shadow of a tree disappears when the sun is directly overhead. In such a special frame of reference, the time component of an object's momenergy—that is, its energy—takes on its minimum possible value, which is equal to the mass itself of that object. However, in whatever free-float frame we view it, the arrow of momenergy spans the same interval, has the same length, possesses the same mass.

Conservation of Momenergy in a Single Collision

The grip of spacetime on mass! That's what we seek to understand. What features of motion does spacetime insist on ruling, and what features does it abandon to the rule of other forces? The answer stands out most clearly in the collision between a moving object, A, and a stationary one, B. In that collision, spacetime gives up total control of the mass of A and the mass of B, and shares its control of motion with electromagnetism, elasticity, and other forces. But like a hardline umpire, spacetime makes sure the collision between A and B is fair.

Let's see how spacetime referees a collision. A mass of 8 units, speeding along at 15/17ths of the speed of light, hits a mass at rest—a mass also of 8 units—and sticks to it. A little figuring shows that the momentum of the speeding object is 15 units, and its energy is 17 units. Because the other object is at rest its energy is identical to its mass, 8 units. After the collision, what momentum, energy, and mass does the resultant object (A:B) have?

Momenergy is conserved; mass isn't. The momentum of the final object is identical to the momentum that the impinging 8-unit mass had before it hit the standing 8-unit mass. That momentum is 15 units, previously the property of A, now the property of the united entity (A:B). Likewise, the energy of the amalgamated object, 25 units, is identical to the total energy that the two objects had before they hit—17 units of energy for the impacting mass plus 8 units of energy for the impacted mass. That's conservation of momenergy in action!

But wait. If the momentum and energy are conserved, then the length of the momenergy vector—the mass—can't be. Indeed, the final object has 20 units of mass, not 16 units. Not on mass as mass does spacetime exert its grip, but on masses (and electromagnetic and other fields) as they contribute to momenergy. Does that grip check up on the total of all masses? No. Enforce conservation upon that total? No. Care about it? No. Does that grip add up the total of all the momenergy? Yes. Enforce conservation upon that total? Yes. Care about it? Yes.

Is there a physical explanation for the increase in total mass from 16 units before the impact to 20 units after the objects hit and stick? Yes,

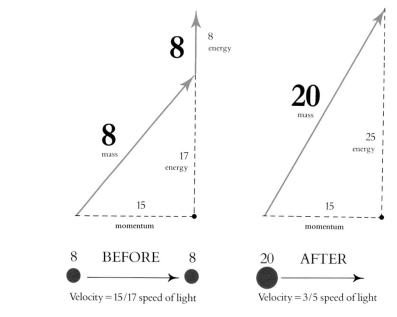

Picture in two-dimensional spacetime of the mass (momenergy), momentum, and energy for two objects and for the amalgamated object formed during their collision.

heat! Energy available in the form of the relative motion of the two objects before they collide and stick goes into energy of internal excitation—or heat—after the two combine. The amalgamated system has more energy than the two objects would if they were tamely juxtaposed. Therefore it has more mass. For the same reason, a hot object has to weigh more than the same object does when cold.

Never yet, however, has anyone succeeded in weighing heat. In 1787, long before Einstein, Count Rumford made valiant attempts to detect and measure the difference in weight between the same barrel of water when hot and when cold; he was unsuccessful. He concluded that "all attempts to discover any effect of heat upon the apparent weight of bodies will be fruitless." Since Einstein, we know that there must be such a difference and can easily figure it. However, the increase in the internal energy of agitation—and therefore in mass—is too small even for today's best instruments to detect. Yet detect it we someday can and must! What a challenge to mankind's ingenuity!

Conservation of Momenergy in Many Collisions

Never has the Pacific seen a man-initiated event that released a greater energy than the Mike nuclear explosion of November 1952. The bomb set off on the Eniwetok Atoll released an energy equivalent to ten million tons of TNT, a thousandfold times the output of the atomic bomb dropped on Hiroshima in August 1945. In the first 100 picoseconds, the first 10^{-7} second, after the uranium, deuterium, tritium, and lithium had completed their reaction, the energy set free resided for the most part right there. It took the form of heat, heat of two kinds. One was heat of agitation—particles zigzagging wildly back and forth between encounters with other particles. The other carrier of heat energy was radiation—particles of light, quanta of electromagnetic radiation, photons. Mike's interior at that instant, at twenty million degrees Celsius or so, was like a barrel of star interior suddenly brought down to Earth.

In that hot mass, during that 10^{-7} second, billions upon billions of photons, electrons and nuclei traded momenergy, each of them billions upon billions of times. Collisions unbelievable in number! Could we have followed one of them? Could we have compared the sum of the momenergy of any two participants before colliding with the sum after their collision? Perhaps. But could we have done that for all the collisions of each participant, and for all the participants? That's beyond the power of any human, any machine, any instrument.

That impossible bookkeeping enterprise spacetime itself nevertheless accomplishes easily, quietly, successfully. It conserves momenergy

Count Rumford (Benjamin Thompson)

Born March 26, 1753, Woburn, Massachusetts. Died August 21, 1814, Auteuil, France. The British physicist and statesman Count Rumford was the first to attempt to weigh heat. Despite numerous other unsuccessful efforts to weigh heat, we know that a given object, hot and cold, must have different masses.

The November 1, 1952, Mike hydrogen bomb explosion that destroyed Eluqelab Islet of Eniwetok Atoll in the Marshall Islands.

as rigorously at the wholesale level during billions and billions of collisions as it does at the retail level during one collision.

How can we picture most vividly this macroscopic conservation of momenergy? Not on any one mass, not even on any colliding pair of masses, do we any longer turn our attention, but on the totality of momenergy carried by all the masses in some standard volume of space (cubic centimeter, cubic meter, or cubic lightyear) and in like standard volumes east, west, north, south, up, and down from it. We do not segment space alone in our bookkeeping. Time, too, we partition into standard segments whose dimensions we measure off by light-travel time (centimeter, meter, or lightyear). Thus we end up with spacetime itself partitioned into blocks. The exact dimensions don't matter. For definiteness, however, picture the block on which we fix our attention as one unit "long," or nearly so, in each space dimension and one unit "high" in the time dimension.

A wild melee of collisions occupies this typical spacetime block. Billions of battles go on, encounters in which one particle gains momenergy, another loses it. But never, in this or any other spacetime block, is any energy or momentum ever created or destroyed. From this single simple *never* statement follows all that need be said, all that can be said, about conservation of momenergy. Does "no creation or destruction of momenergy" in a given region during a given time mean that there's just as much momenergy in a specified region at the end of a specified time interval as there was at its beginning? Not at all! Zillions of particles and photons may flow—and typically do flow—in or out across the boundaries of the region during the time in question. So "no creation or destruction" means—and demands—a more imaginative formulation. Has the bank created any money in my account during the month, or destroyed any, or behaved as it should? To test this point I work out a single simple number. I call it my measure of "creation," although I suspect Mrs. Mykietyn at the bank wouldn't like the name! "Creation" is the amount in the account at the end of the month, minus the amount in it at the beginning, plus the checks paid out in those 30 days, minus the deposits made. Thus formed out of these four numbers, my creation index, were it positive, would indicate creation of money in my account; were it negative, destruction. But always it comes out zero, as the bank and I require. The grip of spacetime on momenergy likewise demands— but also, as we will see in the next chapter, *automatically brings about*—a zero value for an analogous index of creation of momenergy:

$$
\begin{pmatrix} \text{creation of momenergy} \\ \text{in a specified spacetime region} \end{pmatrix}
$$

$$
= \begin{pmatrix} \text{momenergy in specified} \\ \text{region of space at end} \\ \text{of specified time} \end{pmatrix} - \begin{pmatrix} \text{momenergy in specified} \\ \text{region of space at beginning} \\ \text{of specified time} \end{pmatrix}
$$

$$
+ \begin{pmatrix} \text{flow of momenergy out of} \\ \text{left-hand face of region} \\ \text{during specified time} \end{pmatrix} + \begin{pmatrix} \text{flow of momenergy out of} \\ \text{right-hand face of region} \\ \text{during specified time} \end{pmatrix}
$$

$$
+ \begin{pmatrix} \text{flow of momenergy out of} \\ \text{front face of region} \\ \text{during specified time} \end{pmatrix} + \begin{pmatrix} \text{flow of momenergy out of} \\ \text{back space of region} \\ \text{during specified time} \end{pmatrix}
$$

$$
+ \begin{pmatrix} \text{flow of momenergy out of} \\ \text{bottom face of region} \\ \text{during specified time} \end{pmatrix} + \begin{pmatrix} \text{flow of momenergy out of} \\ \text{top face of region} \\ \text{during specified time} \end{pmatrix}
$$

In brief, when any change in the content of momenergy in the region under watch is added to the net outflow of momenergy, the result must be zero.

If the change is positive, then outflow is negative. To say that the net outflow is negative is a fancy but nevertheless useful way to state that the net inflow is positive. That inflow is exactly what brings about the increase of momenergy in the region under consideration. If the change is negative and the momenergy has decreased, then the net outflow is positive. That outflow is what brings about the decrease in momenergy in the specified spacetime block. No "creation" or "destruction" actually occurs. Momenergy is not created or destroyed, it is simply transferred.

An important principle teaches us something new every time we learn to state it in a new way. So here. No creation or destruction of momenergy in a four-dimensional spacetime volume, we now discover, translates itself into information about the *content* of momenergy in eight three-dimensional volumes, eight boundary "faces," eight 3-cubes.

Just as a three-dimensional cube is defined and in a sense "surrounded" by its six two-dimensional faces, so a four-dimensional block of spacetime is defined and "surrounded" by eight three-dimensional cubes. Each of these cubes corresponds to one of the eight terms in the expression above. For example, let's consider two of those cubes: cube 1 defined by the first term, that for the momenergy in the unit region of space at the end of the specified time, and cube 4, defined by the fourth term, that for the flow of momenergy out across the right-hand face of the original 4-cube in the specified time. Cube 4 differs from cube 1 only in the directions of its three dimensions. Cube 1 has three dimensions: left-right, front-back, down-up. For cube 4, however, the dimensions are front-back, down-up, and past-future. Timelike though one of those dimensions is, the cube it helps to define is no less a cube. This cube, like the first, has a content of mass-in-motion for inclusion in our count of momenergy. The amount of momenergy in cube 1 is the amount in the unit region of space at the end of the specified time, while cube 4 contains all the momenergy that has flowed out through the right-hand face of the original 4-cube in the specified time.

The grip of spacetime holds firmly onto the content of momenergy in every one of the 3-cubes that surround a 4-cube. That grip demands a zero *sum* for the contents of all those 3-blocks that bound that four-dimensional spacetime region. In that way the grip of spacetime, there and likewise anywhere, forever bars any creation—or destruction—of momenergy anywhere in spacetime.

A grip on momenergy so vigilant! Does spacetime maintain it by eagle-eyed vision, by clever machinery, by an all-reaching spy system? No, behind all that apparent vigilance lies a linkup of momentum-and-energy to spacetime so simple and so clever, so beautiful and so spare, that it does not count as "machinery" at all! The magic of that grip is now about to show itself in the principle that "the boundary of a boundary is zero!"

7

The Boundary of a Boundary: Where the Action Is!

Leave us now you must —
So you signal to us, Minerva;
But, bringer of wisdom,
Before you go,
Reveal, we beg,
The magic central idea of gravity.

We would venture forth
On many an exploration of our own —
From the harmony of the planets
To the power of the black hole,
And from the clout of a gravity wave
To the dynamics of the cosmos —
But confidently so
Only with your talisman,
Your magic key to it all,
In our hand.

At the boundaries of the box the mass inside grips spacetime outside. Spacetime in turn grips the boundaries and drags the box down upon the scales.

In desperation, we turn to Minerva, the goddess of wisdom, for the key to the magic grip of gravity. Mysteriously she says, "The secret of the grip lies in the boundary of the boundary," and vanishes.

From her enigmatic smile, we know her words somehow divulge the very heart of the mystery. But how can we translate her cryptic message into anything loud and clear?

"The boundary of the boundary," Minerva said. Then we'll start by remembering all that we know about a boundary.

A directed or "oriented" line—a one-dimensional *manifold*—has for its boundary the starting point and the terminal point, both zero-dimensional. Convention counts the end point as positive, as "payoff"; the starting point as negative, as "debt incurred" to start that line. A two-dimensional manifold, a bit of oriented surface cut out of a sheet of paper by a single circuit of the scissors, has for its boundary the directed line—the one-dimensional manifold—traced out by the scissors. And that directed line itself? What is its boundary? Zero! Zero because whatever the point at which we consider that line to have started, that is also the point at which the line terminates. The debt incurred at the starting point annihilates, consumes, eats up the payoff at its end point. Otherwise stated, the *zero*-dimensional boundary of the *one*-dimensional boundary of a *two*-dimensional region is **zero.**

When we change the last sentence a bit, we get another true statement: the *one*-dimensional boundary of the *two*-dimensional boundary of a *three*-dimensional region is **zero.** Really true? Yes, as a friendly dia-

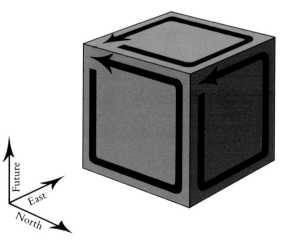

The one-dimensional boundary of the two-dimensional boundary of a three-dimensional space (here a cube) is zero.

gram reminds us. The interior of a cube can be thought of as oriented. That is, each face of the cube inherits from the interior an orientation, a swirl, a direction of circulation as indicated by the arrow that runs around the periphery of that face. All six of the two-dimensional faces together constitute the boundary of the cube.

The one-dimensional boundary of one face of the cube is the directed line indicated by the arrow. Not its start—wherever *that* is—and not its end, because the zero-dimensional start and end cancel. No, the line itself is the boundary. Nothing zero about it, regarded in and by itself. But pick out a segment of that line. Oops! Next to it, on the adjacent face, runs a counter-directed line! Those two line segments annul, kill, annihilate each other. And so it goes at every edge. Result? Total washout of the one-dimensional boundary. The one-dimensional boundary of the two dimensional boundary of a three-dimensional region totals to zero.

Automatic Conservation of Momenergy

Minerva's words, "the boundary of the boundary," have a rhythmic beat. They run through our minds, enchant us, lead us on. Can it be that the *two*-dimensional boundary of the *three*-dimensional boundary of a *four*-dimensional region is likewise self-canceling, self-annihilating, zero?

The phrase "four-dimensional region" sets up vibrations in our memory. Suddenly we remember that in no four-dimensional region anywhere does the grip of spacetime ever permit any change in momenergy whatsoever. No creation or destruction of momenergy in any block of four-dimensional spacetime. And this demand, we discovered in Chapter 6, translates into a rule of regal reach: *content of momenergy totaled for all the 3-cubes that bound any spacetime region must sum to zero.*

Amid the dark mystery of how the grip of gravity works, all at once a flicker of light begins to gleam. Boundary! Our regal rule puts right before our eyes a boundary—the 3-cubes. But where is the *boundary* of the boundary that Minerva mentioned?

Let's turn back, then, and visualize the three-dimensional boundary of a four-dimensional region. It presents to view eight pieces, eight 3-cubes. In the mind's eye one of those 3-cubes begins to gleam all over its two-dimensional faces, gleam like a luminous box of Tiffany glass. Its top glows gold; its front face, red; its right face, blue. Equally vivid colors shine forth on the opposite three faces.

We muse on that many-colored cube, and as we do, our sixth sense picks up the glint of an arriving idea. The glowing faces of the cube start

The 4-D block of spacetime depicted at the center is bounded by eight 3-cubes, shown here as exploded off that block. Each 3-cube has six 2-faces. The content of momenergy—of mass-in-motion—inside each 3-cube expresses itself on its faces. Each face is endowed with an "orientation" or sense of swirl represented here—for a few sample faces—by one-dimensional arrows around its perimeter. Faces with matching colors butt up against each other. Two faces that abut have opposite orientations, opposite senses of swirl, and thus make contributions equal in magnitude but opposite in sign to the audit of momenergy creation in the 4-D block during the stretch of time from "start" (bottom of diagram) to "end" (top of diagram).

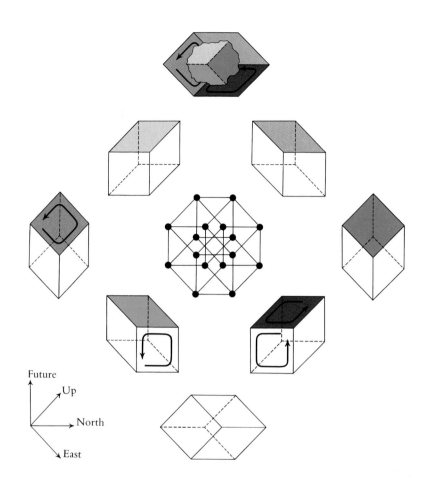

to pulse. We are getting hot on the trail. Does spacetime curvature on the *faces* of the cube grip the content of momenergy inside the cube? In turn, does the content of momenergy inside the cube grip the spacetime curvature on the *faces* of the cube? As if in answer to this reciprocal query, the cube of many colors flashes faster.

Suddenly the light of *the* great idea floods our consciousness. Before we even fully grasp what it is, we know in our bones that the glittering central mechanism of gravity at last lies here, exposed. Our eyes, blinded by the brightness, bit by bit adapt. Outlines become visible before we discern any details.

The two-dimensional boundary of the cube—that's where the action is! That is where the content of momenergy is revealed, and ulti-

Chapter Seven

mately where the grip of gravity grabs. But how can we be sure that the action takes place at that boundary, not inside it? Suddenly a memory springs to my mind. The little fishing town of South Bristol on the Maine seacoast. Farrin's Lobster Pound. A puzzled storekeeper with two identical cardboard boxes before him. He had spent half an hour packing them with styrofoam insulation and cold wet seaweed. As I turned to go, he couldn't remember which one had a single lobster in it, destined for an unmarried friend of mine, and which contained two, as a gift for a young married couple. We had been talking too much. Muttering to himself, he started to unpack one box, thinking to check the contents. His wife stopped him, "Don't unpack, Frank; weigh them!" So he did. He never had to look inside to resolve his dilemma. Weight—a purely outside property—instantly revealed the vital information about what was inside—one lobster or two.

Likewise, never does the grip of spacetime have to reach inside the luminous Tiffany box to sense—and measure—that 3-cube's total content of momenergy. Instead the content of momenergy inside a 3-cube shows up accurately, simply, completely via "scoreboard indicators" on the two-dimensional (2-D) faces of that 3-cube. The sum of those six surface-located scoreboard readings gives the momenergy inside. Outside reveals inside. How, then, do we define and measure the "scoreboard reading" for any one 2-face? Amazingly we don't have to if we're willing to postpone knowing *how much* momenergy lies within the 3-cube and concentrate on checking *conservation* of momenergy. Why? Because any one face—say, the red face—butts up against and cancels out the oppositely swirl-oriented red face of an adjacent 3-cube (see diagram on page 112).

The swirl, the circulation, for each red face—or any 2-D face—is indicated by the arrow that runs around the one-dimensional perimeter of that face. Opposite swirl, opposite sense of circulation—or, in the word of our friends from the world of mathematics—opposite *orientation* means that these circumnavigating arrows run in opposite directions. The two abutting red faces—one belonging to one 3-D cube and the other to an adjacent 3-D cube with an oppositely directed perimeter—are mutually canceling, are mathematical negatives, the one of the other.

The momenergy scoreboard reading that goes with the one red face must be equal in magnitude but opposite in mathematical sign to the momenergy scoreboard reading of the mated red face. Because the momenergy scoreboard readings—whatever *they* may be—that belong to abutting, or mating 2-D faces are opposites, their sum—the momenergy of the interface—is zero. The momenergy scoreboard readings likewise cancel at every other 2-D interface. We feel in our bones that we begin almost to understand how nature conserves momenergy in any four-dimensional block of spacetime. But we also know that no idea is valuable until we can voice it sharp and clear: *conservation of momenergy is automatic because the forty-eight 2-D face-boundaries of the eight 3-D cube-boundaries of any 4-D spacetime region self-annihilate.*

Within a 4-D cube or block of spacetime—a region of space examined for an interval of time—momenergy is automatically conserved. How does nature audit that block of spacetime to make sure that no momenergy is created or destroyed in it? Is it enough for nature to audit the momenergy—the measure of mass in motion—contained in just one of the bounding 3-cubes? No! Instead nature must audit all eight 3-cubes that bound that 4-cube—or, rather, the content of momenergy in all eight of them. Those eight "contents of momenergy" must and do add to zero. That's nature's way to guarantee that never anywhere, in any region of space, studied for any stretch of time, is there ever any creation or destruction of momenergy.

Eight cubes. Six faces for each. That's 48 faces. Each of these two-dimensional faces bears an indicator, a scoreboard, a momenergy register of some yet-unfathomed kind. Not one of those 2-faces stands exposed to any outside world. Everyone of those forty-eight 2-faces abuts another one of them. Twenty-four mated pairs!

Whatever the scoreboard of momenergy carried by any one 2-face, the scoreboard of momenergy carried by its mated face has to, and does, read the exactly opposite number. And so it is for the six faces of the next 3-cube, and the next 3-cube, and the next, all the way around all eight 3-cubes that bound that 4-D region, that block of spacetime, that audited domain, in which nature demands zero change of momenergy. This conservation condition fulfills itself automatically. How? By the principle that the 2-boundary of the 3-boundary of a 4-region is zero. That is how nature ensures and guarantees conservation of momenergy.

Action of Mass-in-Motion Inside on Spacetime Outside

Bravo! The content of momenergy inside a 3-cube, we conclude, must somehow reveal itself—without any internal probing whatsoever—through the sum of scoreboard readings on the six 2-D faces of that 3-cube. But reveal itself how? Make itself felt how? Suddenly we remember from our earlier discussions (Chapter 6) that the content, the magnitude, the measure of momenergy is mass. "Momenergy make itself felt how?" is the same as "Mass make itself felt how?" The answer to the second question and hence the first we discovered in Chapter 5: Mass—most vividly, the mass of Earth itself—makes itself felt in the curvature that it imposes on spacetime. Curvature manifested not in the free-float motion of one mass, but in the relative motion of two geodesics, two histories of free-float motion, two nearby and nearly parallel worldlines of two microscopic test masses whose directions of travel slowly bend—or twist—or bend *and* twist—relative to each other. Two such geodesics can also define for us two of the opposite edges of a 2-D face, which though nearly parallel cannot be completely so, because of the slight bending of the worldlines. This bending, this rotation of two geodesic boundaries of that face relative to each other, does it not provide at last our long-sought 2-face "scoreboard indicator" for momenergy? Do we indeed get a true count of the momentum and energy contained within a 3-D cube by adding the angles of bending associated with the six 2-D faces of that cube? What could be simpler? We have only to ask this question to feel ourselves happily launched on the wonderful way to understanding the grip of mass-in-motion on spacetime.

We can measure the bending, the angle, the rotation associated with each 2-D face, or plaquette—a two-dimensional slice of spacetime—by the method of parallel transport that we investigated in Chapter 5 (page 79). Geodesics, free-float worldlines, run along and define two nearly

The rotation, *or bending-plus-twist, associated with the 2-D face of a 3-D cube, one of the eight blocks that bound a region of 4-D spacetime. The sweeping green and blue lines mark two of the edges of the face. For simplicity we envisage both as free-float worldlines that start parallel at the remote end of the face—and think of Blue as carrying a radar mounted on a gyroscope-stabilized platform. To "start parallel" means that initially Green shows up on Blue's radar as having a zero speed of approach and a zero rate of change of direction. In consequence of the curvature of the intervening spacetime, by the end of the run Green shows up as not only approaching but also changing direction. In other words, spacetime curvature has rotated Green's spacetime velocity relative to Blue's. This rotation has been translated, in the inset at the center of the 2-D face, to two arrows of unit length and a parallelogram, which they span and thereby define. The "orientation" or sense of swirl (gold arrows) has this in common with the parallelogram and the 2-D face, that it runs concurrent with the edge or arrow that belongs to Blue.*

parallel edges of each face; the other two nearly parallel edges cut those two free-float worldlines. Thus they define the start and stop of the stretch of line in which we are interested. Moreover, each edge of any one face also is an edge for an adjacent face. Belonging to one face, that edge is traversed in parallel transport in one direction. Belonging to the other face, it is traversed a second time but in the opposite direction. The first traverse contributes to the rotation associated with the one face. The second traverse makes a contribution to the rotation of the other face equal in magnitude but opposite in sign to the rotation associated with the first face. Thus when we add all six rotations, we count the contributions of all the edges twice over—once with a plus sign and once with a minus sign. The total? Zero! Zero for our intended measure of the momenergy inside the 3-cube! Heavens! Something *must* be wrong in our thinking.

We stumble but continue. Any minute now, we're sure, we're going to discover the answer to the central question: how does mass-in-motion *inside* reveal itself in spacetime bending *outside,* at the surface? The many-colored faces of the Tiffany cube flash encouragement.

In imagination we see a shack adrift upon the ice of a frozen sea. The owner is making an urgent appeal: turn the building to face south but don't displace its center! Is this possible? Two men appear and start pushing on the shack with equal and opposite forces. No displacement of the center occurs, but the lodge slowly starts to turn. Why? The separation between the lines of action of those two forces, a perpendicularly measured distance between them, acts as a "lever arm." The length of that lever arm multiplied by the force gives what students of mechanics recognize as the *turning moment,* or *torque* responsible for that turning. Net force? Zero. Net turning moment of those forces? Non-zero!

As in mechanics, so in geometrodynamics. In mechanics, the central idea is force. In geometrodynamics it is bending or rotation—the rotation of the direction of travel of one free-float world line, one geodesic, relative to a nearby and nearly parallel free-float geodesic. Here we are concerned with the rotation associated with one of the six 2-D faces that enclose a little 3-D cube. In mechanics we move from a force to the moment of that force in a single step by multiplying the force by the distance from a fulcrum (i.e., axis of rotation) to the line of action of that force. Here, too, in one leap, we move from a rotation to the moment of that rotation by multiplying the rotation by the distance from the fulcrum to the center of the 2-face associated with that rotation.

In figuring the grip of momenergy—of mass-in-motion—on spacetime, it does not matter where, in imagination, we place the fulcrum. How come? As well ask why a new choice of fulcrum does not change the value of the total torque exerted by two equal forces upon the shack. In the case of the shack, the sum of the *forces*—due allowance being made for their counterdirectedness—is zero. Here, analogously, the sum of the *rotations* associated with the six faces of a 3-D cube is zero.

The tempo of the flashing colored faces has risen. The long-sought answer to the action of mass inside on spacetime outside, it signals, now lies in ready reach. In the turning of the shack, the key quantity was not the force—for the two countervailing forces added up to zero—but the moment of that force. Likewise, in the action of mass on spacetime, the vital quantity associated with a 2-D face is not the bending, not the angle, not the rotation, but the *moment of rotation*. The rotations, summed up for the six faces, add to zero. But the moments of rotation don't! *The sum of the moments of surface-located rotation reveals and measures the amount of mass-in-motion—the momenergy—inside the 3-cube:*

$$\begin{pmatrix} \text{sum of moments} \\ \text{of rotation for} \\ \text{the faces of a} \\ \text{little 3-cube} \end{pmatrix} = 8\pi \times \begin{pmatrix} \text{amount of} \\ \text{momenergy within} \\ \text{that 3-cube} \end{pmatrix}$$

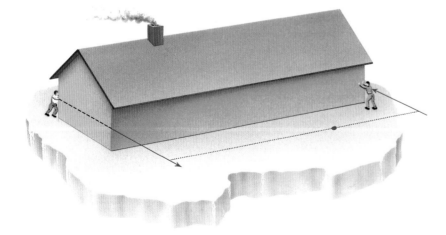

The total turning moment is independent of the location of the fulcrum. *Two men push on the opposite corners of the shack with equal but opposite forces. Each exerts a torque or turning moment about a fulcrum located in imagination wherever the beholder chooses to place it. The perpendicular distance from that fulcrum to either force governs the moment which the beholder attributes to that force. However, the sum of the turning moments, the total torque, which is what counts for swinging the shack around, does not depend on the placement of the fulcrum.*

This single simple expression—the Einstein-Cartan equation—gives us the most vivid image that mankind has ever won of the living heart of gravity. Here shines forth the influence of energy and momentum—whether of mass-in-motion, or of electromagnetic fields, or of other fields, or of all of these—on spacetime curvature at the two-dimensional boundary of any elementary 3-D region. Here glitters in a simple geometric form Einstein's 1915 battle-tested and still-standard *law of geometrodynamics,* his famous field equation. This equation holds the answers to every question about gravity that we know enough to ask.

A CLOSER LOOK AT MOMENT OF ROTATION

The moments of rotation associated with the top and bottom faces of a 3-cube are represented as parallelepipeds, three-dimensional objects. One edge of each is the line from a fulcrum, or center of rotation, inside or outside the 3-cube to the center of the face in question. The side of each parallelepiped away from the fulcrum is a little parallelogram. Spanning the parallelograms are two arrows of *unit* length. They symbolize the initial and the final orientation of a marker arrow to be carried, in imagination, in parallel transport around the perimeter of the face in question to measure the rotation associated with that face. The volume contained within the parallelepiped expresses itself, then, not in cubic meters but in meters—that is, (meters) × (dimension-free number) × (dimension-free number). And the meter, too, is the measure of mass, and of momentum and energy. As far as units of measure go, we are evidently on the right track. But momenergy has the quality of an arrow, an arrow that points in a well-defined direction in spacetime. Whoever heard of a 3-D parallelepiped pointing anywhere? Our everyday intuitions, however, unconsciously assume that the objects we examine sit in a 3-D space. In spacetime, there *is* a well-defined direction perpendicular to the 3-D parallelepiped. *That* is the direction of that parallelepiped's contribution to the arrow of momenergy! This feature of the grip of gravity is like the direction of a torque—a turning moment—such as the direction in which a screw advances through the wood when we twist it to the right! With direction now evident for each face's contribution to the momenergy, and its magnitude also evident, six directed arrows stand forth and have only to be combined, as arrows combine, to reveal the total momenergy contained in the 3-D cube.

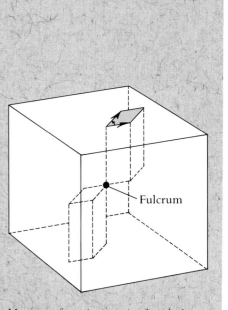

Fulcrum

Moments of rotation associated with the top and bottom faces of a 3-cube, represented as parallelepipeds. The sum of the volumes of the parallelepipeds associated with all six faces tells us how much momenergy there is in the 3-cube. That sum is independent of the location of the fulcrum (black dot). The perpendicular—in 4-D spacetime—to that totalized 3-volume determines the direction of the arrow of momenergy.

Geometrodynamic Field Equation

Einstein's field equation can tell us the gravity—that is, the spacetime curvature—in and around a star, a black hole, or any other spherically symmetric center of mass. And it can confirm the strength and directional properties for the tide-driving curvature outside that mass, which experience already tells us are correct (page 89).

Can it tell us in a jiffy the story of the expansion of a closed model universe, and of the slowing of this expansion, when the model is so simple that in it the density of mass departs nowhere greatly—on the average—from uniformity? Yes.

Can it tell us the workings of a gravity wave—how the collapse of a distant star generates it, how it zings through space, how it makes the Earth tremble beneath our feet? Yes.

All this and more follows from the geometrodynamic field equation. This marvelous statement tells how mass—how momentum and energy—grip spacetime. It reveals how mass, there, bends spacetime geometry, there—deforms it, warps it—as a jumper warps the trampoline beneath his feet. Cupping of the fabric there, however, demands and enforces warping on the canvas in all the domain roundabout. Likewise spacetime geometry there, bent in one way where mass sits or moves, demands and enforces bending of another kind on all the surrounding empty spacetime.

Bending spreads its influence from region to region. The Einstein geometrodynamic influence equation describes and quantifies this spread of influence from region to region. In any region of emptiness, momenergy is zero. So is the moment of rotation—because zero momenergy in any locale implies and demands that there the moment of rotation also must be zero. But for the shack sitting on the ice, with two men pushing on it at different places and in opposite directions, does a zero total turning effect, a zero sum for the separate turning moments of those forces, mean that those forces themselves are individually zero? Not at all.

As in mechanics, so in geometrodynamics! In the emptiness outside a center of mass do we find a zero disturbance, a zero measure of geometric influence, a zero departure of spacetime from ideal? First, the moment of rotation. Yes, that's zero. Second, rotation itself, relative bending of nearby and nearly parallel free-float worldlines, change in their relative motion. Like forces in mechanics, the rotation is not zero. Were it zero, there'd be zero gravity outside the Earth, no planetary orbits, no gravity waves! That rotation, that bending, that change in the relative speed of nearby worldlines—that's what gravity, examined *locally, is!*

Élie Cartan

The French mathematician and geometer Élie Cartan (1869–1951) used his calculus of "exterior differential forms" to derive the simple Einstein-Cartan equation from Einstein's much more complicated geometrodynamic field equation. Cartan's boundary-of-a-boundary geometric approach to Einstein's theory provides a wonderful simplicity. In 1929, Cartan sent Einstein some of his mathematical results concerning general relativity. Einstein wrote in reply: "I didn't at all understand the explanations you gave me . . . ; still less was it clear to me how they might have been useful for physical theory." Einstein's difficulties may have stemmed from Cartan's "extremely elliptic style that . . . has baffled two generations of mathematicians," according to Jean Dieudonné.

Einstein's influence law thus displays not only the grip of mass where it is on spacetime where it is. It also tells how bent spacetime here grabs and bends adjacent spacetime.

Even beyond the grip of mass on spacetime, the moment-of-rotation equation reveals and rules the grip of spacetime on mass. Grip *on* mass? Yes, reaction to the grip *of* mass! That's the other half of the story of gravity. That grip on mass enforces the law of conservation of momenergy. Enforces it automatically. Enforces this conservation in a subtle, clever, hidden way. Enforces it via the principle that the 2-boundary of a 3-boundary of a 4-D spacetime region is zero!

We have traveled an adventurous course in these last pages. Content of momenergy *within* a 3-cube, we have found, is the sum of moments of rotation—rotation, or relative bending of worldlines, or spacetime curvature—as evidenced at the six 2-D faces of that 3-cube. Every single one of these 2-D faces, however, is paired with a totally compensating face when we consider not a single 3-D cube but all eight of the 3-D cubes that surround an elementary 4-D block of spacetime. In consequence of this pairing, this mutual annihilation of these *faces,* the sum of the momenergy contained *within* all eight 3-cubes adds up to zero. Zero total content of momentum and energy as summed up for the eight cubes that constitute the *3-D boundary of that little spacetime region.* And that's our sophisticated way to say that in that region there is total balance between what goes in and what goes out! By extension to other times and other places, this central boundary-of-a-boundary feature of geometrodynamics ensures that never at any time or any place is there ever any creation or destruction of momenergy whatsoever!

Is Physics at Bottom "Law without Law"?

This great world around us—how is it put together? Out of gears and pinions? By a corps of Swiss watchmakers? According to some multifaceted master plan embodying an all-embracing corpus of laws and regulations? Or the direct opposite? Are we destined to find that every law of physics, pushed to the extreme of experimental test, is statistical—as heat is—not mathematically perfect and precise? Is physics in the end "law without law," the very epitome of austerity?

Nothing seems at first sight to conflict more violently with austerity than all the beautiful structure of the three great field theories of our age—electrodynamics, geometrodynamics, and string-theory dynamics. They are the fruit of a century of labor, hundreds of experiments, scores of gifted investigators. How can we possibly imagine all this ordered richness originating in austerity? Only a principle of organization that is no organization at all would seem to meet any demand for total austerity. In all of mathematics, nothing of this kind more obviously offers itself than the principle that the boundary of a boundary is zero. Moreover, this principle occupies a central place in all three of today's great field theories. To this extent almost all of physics founds itself on almost nothing.

Far-seeing Gottfried Wilhelm Leibniz advocated a still greater vision of existence: "For deriving everything out of nothing one principle suffices." Was he right? Underneath the workings of the world will someday a humble, thoughtful, gifted knot of searchers lay open to view the great unifying principle? Noble work for man! Rich gift to mankind!

To confront issues so cosmic is to turn back with a fresh eye to how gravity works—via the double grip of spacetime on mass and of mass on spacetime. Nowhere better than by examining that grip can we see more vividly, more instructively, and more impressively the reach and power of that austerity-flavored principle, *the boundary of a boundary is zero*. And nowhere will we find that principle operating more beautifully, more simply, and with more direct ties to everyday experience than in the warping of space—and of spacetime—around a spherically symmetric center of attraction.

8

From Potter's Wheel to Space Geometry

Oh potter, shaping with your hand
That clay that turns upon your wheel
Tell us that you fully
Fix the form of the pot
By your choice of profile.
Instruct us how to choose that profile —
When we turn out space geometry
Upon the potter's wheel
Of our imagination —
So that its space curvature
At every point
Outside Earth, Sun, or black hole
Comports with total emptiness,
With zero moment of rotation,
With zero density of mass.

Earth's shape exemplifies the near-sphericity enforced on mass by gravity.

Karl Schwarzschild

Born October 9, 1873, Frankfurt am Main, Germany. Died May 11, 1916, Potsdam, Germany. Schwarzschild was a pioneer in devising instruments, making measurements, and applying physics to understand the working of the stars. While serving as artillery lieutenant in the German army on the Russian front, and shortly before his death, he worked out the first exact solutions of Einstein's geometrodynamic field equation, key to every simple picture of spacetime geometry around a center of gravitiational attraction.

No feature of gravity so much simplifies the study of gravity as its pulling of matter into those spherical collections of mass that we call planets, stars, and black holes. The coalescing material, full of turbulence and eddies, endowed with huge bulges and protuberances, twisting and vibrating, settles down into a condition that—unless endowed with extraordinary spin—is essentially static and spherically symmetric.

As matter coalesces into a sphere, the violence of the adjustment generates waves in space geometry. They are richer in energy than the ocean wave set up when two giant freighters crash head-on in the fog and lock together into one. Some of these *gravity waves*, or *waves in spacetime geometry*, carry energy outward, away from the new star. Some of them carry energy inward, into it.

Far simpler than the story of these waves, however, is the message of gravity about conditions *after* the shakedown to sphericity. This message—the gold of gravity—is the curved spacetime geometry that surrounds Earth and Sun, neutron star and black hole, and every other center of gravitational attraction that is spherically symmetric or essentially so. Karl Schwarzschild, in 1916, first unearthed this message in an exploration of the consequences of Einstein's geometric theory of gravity. That is why spacetime geometry around a sphere is now generally called *Schwarzschild geometry*. Another name for it is *black-hole geometry*. Whoever has that gold in hand can buy an understanding of almost every other feature of gravity.

The importance of Schwarzschild geometry derives from the tendency of matter to agglomerate into masses that are spherical or nearly so. Its simplicity derives from its static character and its spherical symmetry. Not a ripple, wave, bulge, or vibration disturbs its perfection. One parameter, and one alone, suffices to fix the Schwarzschild geometry in its entirety in the region outside a central object. This solitary autocrat that rules the Schwarzschild geometry is the magnitude in geometric units of the central mass: 0.44 centimeters for the Earth, 1.477 kilometers for the Sun, and even more for a black hole.

No feature of the spacetime around a static, spherically symmetric center of attraction is more striking than the "space part" of this Schwarzschild geometry. I like to compare the curving of this geometry outside the object with the curving of the fabric of a trampoline outside the zone of impact of a jumper. Inside that zone, the placement of the jumper's feet and toes matters; outside, it doesn't. Likewise the placement of mass inside the Sun, Earth, or other attracting object determines the special form that the geometry takes inside. Not so outside. Outside, all that matters for the form, the shape, the curvature of the geometry is the total mass inside.

Chapter Eight

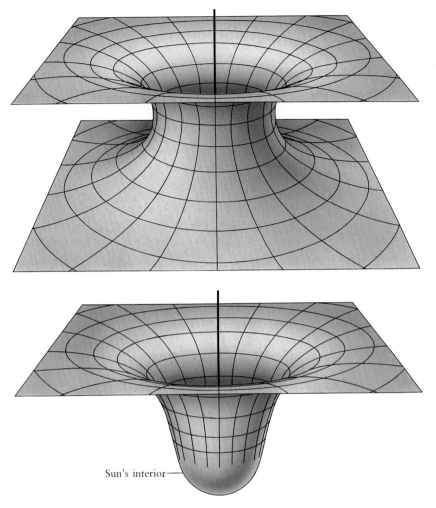

Sun's interior

Rotational Symmetry and Embedding

There is no simpler way to grasp the space part of the Schwarzschild geometry than to compare it with the geometry around the throat of a water jar being shaped on a potter's wheel. The water jar, formed as it is by turning, has an *axis of rotational symmetry*. This axis runs down the middle of the jar. Yet it is not—except at the bottom—a part of the jar. It is a part of our understanding of the shape of the jar. We can visualize and understand that shape from a single glance at a picture of the jar. How do

In picturing the shape of the two-dimensional surface of a pottery jar, we unconsciously embed it in three-dimensional flat space. By analogy, we can embed the three-dimensional curved space that surrounds a center of attraction in a flat four-dimensional space.

we do that? The axis helps. But the mind's customary picture-forming tool lies deeper, so deep that we use it almost unawares. We stop thinking of the smooth face of the jar as sitting in the splendid isolation of a two-dimensional existence. Instead, we envelop that surface in imagination with three-dimensional empty space. We embed the 2-D surface—in the mind's eye—in the 3-D space that Euclid wrote about. Embed it in 3-D space of the kind envisaged by the everyday land surveyor. Embed it in 3-D space to which the potter, working at her wheel, would be crazy to attribute any curvature.

To embed—in imagination—is not a necessity. It is an option. We could choose not to exercise that option. We could seek to learn about the curvature of the 2-D surface all that ants can ever know, ants whose existence is confined to it. We could figure out the curvature in any locale by the methods of bending, parallel transport, or distance as applied right on the surface (Chapter 5). However, for us it is simpler to visualize the curvature of the surface of the throat of the jar by considering it to be embedded in an enveloping three-dimensional space. For any simple picture—whether of a potter's jar or of the space part of the Schwarzschild geometry—embedding as a surface of revolution is key.

One important difference between a jar and space geometry, however, has to be confronted. The inner curved surface of a jar is two-dimensional, but the space part of the Schwarzschild geometry is three-dimensional. The jar's surface has only one axis of rotational symmetry. In contrast, Schwarzschild space, like any geometric object endowed with spherical symmetry, has an infinity of axes of rotational symmetry. How can we come to terms with this multitude of axes of symmetry and with the one-higher dimensionality of the Schwarzschild space geometry compared with the jar's surface geometry? By adding one more dimension to the flat Euclidean space in which we imagine the curved-space geometry to be embedded!

Do we have difficulty picturing objects in this new and wider framework of "four-dimensional Euclidean space" with a "curved three-dimensional space geometry" embedded in it? Then we can cheat! We can privately picture a geometry lower in dimensionality by one unit—that is, flat everyday three-dimensional space with a two-dimensional surface of revolution embedded in it. Then we can talk of this two-dimensional surface as if it were the three-dimensional curved space part of Schwarzschild geometry. Talk of the three-dimensional space in which it is embedded as if it were four-dimensional. And talk of the one axis of rotational symmetry pointing in one direction as if it were an infinity of axes of rotational symmetry, passing through a common point and pointing in all directions. With such a judicious combination of thought, talk, and picture, we are in a fair way to grasping the main features of the curved space geometry that surrounds a planet, star, or black hole.

One caution is appropriate. It concerns the flat Euclidean space of four dimensions in which curved three-dimensional space is pictured as embedded. *This* fourth dimension, this "embedding" dimension, has nothing whatsoever to do with time. To reach off the curved 3-geometry into that imagined four-dimensional space is totally impossible. There isn't any real physical space there for us to reach into. For an ant, running around on the inner surface of the jar, and climbing on the back of another ant, there is!

The shape of the Schwarzschild curved space geometry follows a remarkably simple curve, discoverable by astonishingly simple reasoning. It is generated by revolving a *parabola* about its *directrix*, another name for the axis of revolution. By this means is created a curved sheet called a paraboloid. The distance from the axis of revolution to the nearest point on the paraboloid is equal to twice the geometric measure of the mass of the center of attraction: that is, twice 0.44 centimeters or 0.88 centimeters for the Earth and twice 1.47 kilometers or 2.94 kilometers for the Sun. This, in brief, is the story of the space part of the Schwarzschild geometry.

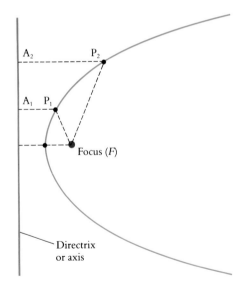

A parabola (green curve) is the locus of points every one of which is equidistant from the focus (F) and the directrix (vertical axis). Thus, $A_1P_1 = P_1F$ and $A_2P_2 = P_2F$.

Curvatures Balance in Mass-Free Space

By what magic influence does space outside a static spherically symmetric center of mass bend itself into a paraboloid—a paraboloid endowed with infinitely many axes of rotation, of which only one shows in our one-lower-dimension potter's wheel picture? To understand that "magic," imagine a point in space surrounded by an imaginary cube completely devoid of mass, of energy, of momenergy. As we discovered in Chapter 7, the momenergy within a small cube anywhere in spacetime is directly proportional to the sum of the moments of rotation associated with its faces. For a little cube of space, empty of momenergy, the moments of rotation must add to zero. In consequence, curvatures on the faces of this space cube have to compensate; in other words, the curva-

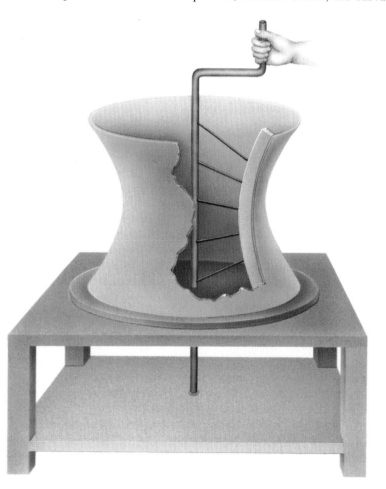

Generating a paraboloidally curved 2-geometry. This is a fancy name for a bottomless pot formed upon a potter's wheel by a cutting edge with the shape of a parabola. It is attached by rods to the directrix of that parabola—the stationary rod that lies along the axis of the potter's wheel. Alternatively, the clay can sit stationary and the parabola can be cranked around to generate the paraboloid— the usual way of speaking in mathematics.

tures must add to zero. This compensation of curvatures, as we will see, is possible only when the cube exists in a space whose shape follows the curve generated by a parabola. The parabola, rotated around "the" axis (actually, many axes), generates Schwarzschild geometry.

The condition of zero moment of rotation is the great and governing principle behind Schwarzschild geometry. It rules the compensation of curvatures and selects the parabola as generating curve for the space part of spacetime. To see why this is so, let's consider a point in space and build around it in imagination a tiny cube, bounded by six plaquettes— three pairs of parallel faces. One pair of faces runs parallel to the surface of the spherically symmetric attracting mass, which lies outside the cube. I'll call them Sun-facing faces or, briefer and better, *exposed* faces. Of the other four faces, each has two of its edges also running parallel to the

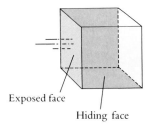

Exposed face

Hiding face

Any locale in mass-free spacetime can be pictured as caged-in by a cube bounded by six faces, or plaquettes. The two exposed faces (yellow) are oriented head-on to the attracting mass; the other four faces, the hiding faces (gray), are oriented edgewise. Because there is no mass, no momenergy in the space cube, the sum of the moments of rotation associated with all six plaquettes must equal zero.

surface of the Sun or other attracting mass. The other two edges of such a face, however, run straight out along the line from the center of attraction. The face stands edgewise to the Sun and thus gets minimal illumination. I'll call such a face a *hiding* face. There are four of them. Thus two exposed plaquettes and four hiding plaquettes together frame the little cube.

Let's select one of these faces and examine it more closely. It has four edges. Each edge is a geodesic. Being a geodesic, it runs as straight as it possibly can through the curved space that encompasses our imaginary cube. As a consequence of that curvature, two opposite edges that start off parallel don't stay parallel. They bend toward each other, or twist relative to each other, or away from each other. In this sense they undergo a relative rotation.

The amount of the relative rotation of direction—expressed as an angle and measured in radians—is equal to the product of two quantities. One quantity is the area of the plaquette encompassed between the two geodesics after any given length of travel. The other quantity is the space curvature associated with this plaquette. Conclusion? With each of the six plaquettes that bound our imaginary cube of space are associated three quantities: its area, its curvature, and—the product of these two—the relative rotation of direction of either one of the two pairs of geodesics that bound the plaquette.

Now we turn from rotation to moment of rotation. That moment is the heart of the grip of spacetime on mass and, when there's no mass present, the grip of spacetime on *spacetime*. As we reasoned in Chapter 7, by analogy to force in mechanics, the moment of rotation associated with a plaquette equals the distance from a fulcrum to that plaquette multiplied by the rotation associated with that plaquette. Where we locate the fulcrum influences the magnitude of the moment of rotation that we attribute to a given face. The location of the fulcrum, however, has no influence on the sum of the moments of rotation as figured for all six faces. Having total freedom in where we'll put the fulcrum, we let simplicity be our guide and locate the fulcrum at one corner of our space cube. This point, lying at a place where three faces intersect, has zero distance from any of them. Thus the moment of rotation associated with these three faces is zero, and so, obviously is their sum.

Only three plaquettes remain to be considered. The moment of rotation of each is given by

$$\begin{pmatrix} \text{moment of rotation} \\ \text{of face} \end{pmatrix} = \text{distance to fulcrum} \times \text{rotation}$$

$$= \text{distance to fulcrum} \times \text{curvature} \times \text{area of face}$$

The product of the first factor (distance) and the last factor (area) is simply the volume of our little space cube. This volume, clearly a nonzero quantity, contributes an identical factor to the moment of rotation of each of the three remaining faces (one exposed face and two hiding faces). The only way for the moments of rotation of these faces to add to zero is if the middle factor, the curvature factor, for the three faces adds to zero:

curvature on exposed face + (2 × curvature on hiding face) = 0

In brief, curvature on an exposed face must be twice as big as, and opposite in sign to, the curvature on a hiding face. Here we have the principle of the balance of curvatures falling straight out of the principle that the moment of rotation is zero in any region where there is no momenergy.

We arrived at this conclusion without appealing to any ideas about embedding in a flat space of one higher dimension, or rotation of a parabola—or any other generating curve—about a directrix or axis of rotational symmetry. However, when we *do* choose to use the ideas and language of embedding, we can define two quantities associated with the curvature:

■ *Rafter:* the distance from a given locale on the generating curve—measured perpendicularly off it—to the directrix or axis of rotation.

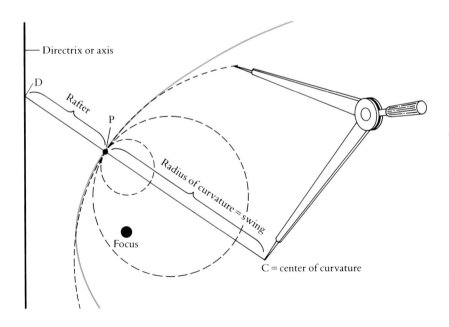

To find the radius of curvature at any chosen point of a parabola, first erect a perpendicular at that point. On that perpendicular, pick some trial point and there dig in the sharp point of a compass. With the other leg of the compass, the one that bears the pencil, start at the chosen point of the parabola and trace out a circle. If the circle hugs the parabola in the chosen locale, then the circle's radius is identical with the radius of curvature of the parabola. The radius of curvature is also called the swing; its value—for the parabola, and only for the parabola—is twice that of the rafter, the perpendicular distance from the given locale to the axis.

- *Swing:* the radius of curvature of the generating curve in the given locale; that is, the length of string we need to swing a circular arc that will hug the generating curve in this locale.

In terms of these two distances—which only make sense when we adopt the language of embedding and generating curve—the principle of the balance of curvatures in a mass-free region takes the form

$$\frac{1}{(\text{rafter})^2} - \frac{2}{\text{rafter} \times \text{swing}} = 0$$

or, equivalently,

$$\text{swing} = 2 \times \text{rafter}$$

Only the parabola, of every conceivable generating curve, has the property that its swing is twice its rafter. Thus, as a consequence of the condition of zero moment of rotation, curved space geometry must have a paraboloidal shape.

Soap-Film and Curved Space

The shape of a soap film, which also is governed by a simple principle, can help us understand more about Schwarzschild geometry. At each point on a soap film there are two principal curvatures. Those curvatures have to be equal and opposite for any free-standing film. Otherwise the tension in the film, the molecule-to-molecule attractions, will pull the film more powerfully to one side or the other. That is why the two principal curvatures, varying though they do from point to point, nevertheless at each point individually are equal in magnitude and opposite in direction.

When two identical metal loops are held parallel at a modest separation, dipped into a soap solution, and carefully pulled out, with good luck, or good maneuvering, or both, a soap film will form which joins the rings but does not bridge either ring individually. Halfway between the one ring and the other is the waist. At a point on this waist, it is easier than it is anywhere else to examine the two principal curvatures of the surface and verify that they are equal in magnitude but opposite in direction. When we compare this soap film with the section of Schwarzschild geometry depicted earlier (page 125), we see some similarity: a waist, and at this waist—and everywhere—possible directions of travel that curve

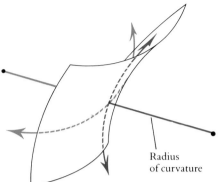

Left: *Soap film obtained by dipping two closely spaced wire loops into a soap solution and then withdrawing them.* Above: *Exerted on each bit of the soap film by the surrounding film (not shown) are tensions. The up and down tensions (red arrows) do not quite compensate because they do not have quite opposite directions. Together they give a net pull to the right, which is proportional to the curvature of the surface along the dashed red line. Likewise, the two tensions shown in blue together give a net pull to the left. It is proportional to the curvature of the surface along the dashed blue line. The soap film adjusts its shape until the two radii of curvature, (red and blue solid lines) are equal in magnitude and opposite in direction, as indicated.*

Radius
of curvature

out, but also possible directions of travel that curve in; in other words, oppositely directed linear curvatures. The surface curvatures of paraboloid and soap film are also qualitatively similar. They are not positive—that is, sphere like. They are negative—that is, saddlelike.

Closer consideration, however, reveals a significant difference. Everywhere on the soap film the radius of curvature that points out is identical in length with the radius of curvature that points in. In contrast, everywhere on the paraboloid of the Schwarzschild geometry, the radius of curvature that points out—the swing—is twice as big as the rafter—the radius of curvature that points in.

There is a still more important difference between the two geometries. One, the soap film, is a two-dimensional surface. The other, the Schwarzschild or black-hole geometry, is a three-dimensional curved space. Pictured as embedded in an enveloping flat or Euclidean four-dimensional space, it has at each point *three* principal radii of curvature, not two.

The most important difference between soap film and Schwarz-schild space geometry, however, lies in the kind of curvature that governs each: for the film, it is the balanced curvature of two one-dimensional lines; for the space, it is the balanced curvature of six two-dimensional surfaces. What do these terms mean? And how are they related? To explore these differences a little is to arrive at new insights on what curvature is and how it works. To do so, we do not have to ascend to the traditional "metric," tensor calculus, partial differential equations, and other mathematical mountains familiar to analysts of Einstein's field equation. We will follow a new route. We will look at the curvature of space directly, graphically, geometrically.

The Two Principal Radii of Curvature

The quickest way to appreciate the three principal curvatures of the Schwarzschild 3-geometry is to look at the two principal curvatures of the inner surface of a bottomless water jar after it has been fired and has attained the hardness of pottery. To start, let's drop the jar on the floor so that it smashes into a hundred pieces. To examine the curvature of one of these shards, we will use a multitude of little metal arcs cut out of a sheet of brass with scissors. For some, the edge has a concave curve; for others, the curve is convex. The edge of each template has its own radius of curvature. One by one we try fitting them to the smooth thin bit of jar. We soon come to appreciate that there are two extremes of curvature on the chosen shard. The one extreme is attained only when the metal template runs in one direction. The other extreme of curvature is found only for a template that runs in a different and perpendicular direction. Directions between these two "principal directions" give curvatures between these two extremes. A standard theorem of surface geometry tells us that we have only to know the two principal curvatures, and the direction of one of them, to predict the 90-degree-away direction of the other—and to predict the curvature for every intermediate direction.

Instead of speaking about curvature, we can equally well speak about radius of curvature. The smallest radius of curvature means the biggest curvature; the biggest radius of curvature means the smallest curvature. With a lump of plastic adhesive we attach to the shard two superlong matchsticks. One is cut to the right length and pointed in the right direction to represent one of the two principal radii of curvature. The other matchstick agrees in length with the other principal radius of

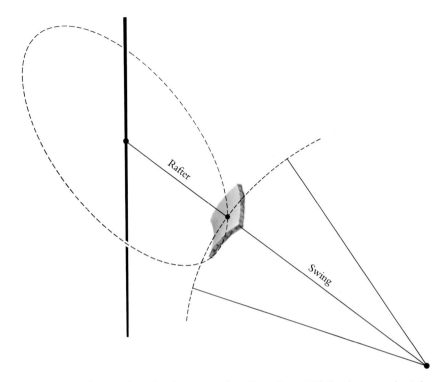

curvature, but points in the opposite direction. While the matchstick length shows the magnitude of a principal radius of curvature, the matchstick tip reveals the location of the center of that principal curvature.

We now reassemble the fragments, with the help of glue and patience, into the original pot. As we do, a remarkable circumstance catches our attention. The tip of the inward-pointing matchstick lies on the axis of rotational symmetry of the jar. The direction in the surface associated with this principal curvature is identical, moreover, with the direction of scraping of the potter's tool—and reasonably enough. What other preferred direction is there at that point on the surface of the jar?

Are we happy that one of our so carefully located matchstick tips—one of the two centers of curvature of our shard—turns out to lie on the axis of revolution? Yes, but concerned too—concerned whether the location of the tip on the axis is not unreasonable. After all, as the clay is being formed on the potter's wheel, isn't every point of it being carried round and round in a circle? And isn't the center of that circle, isn't *that* point, the center of curvature? Located on the axis though it is, it lies at a point on that axis very different from where the tip of the matchstick touches it. Which then is right? Center of circle or matchstick tip? The center of the circle could only define the principal radius of curvature for

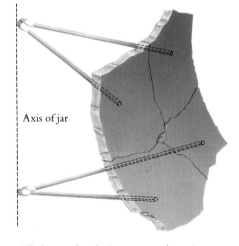

Axis of jar

All the matchstick tips meet at the axis.

the shard in question if that bit of clay were to have been tipped up vertically as it was being scraped, formed, turned. But it wasn't. It had a tilt. So does the matchstick that stands perpendicular to it. That's how it comes about that the relevant center of curvature of the little bit of surface, while lying on the axis of symmetry, also lies on the perpendicular to the jar.

What about the outward-pointing radius of curvature? For clay being turned on a potter's wheel, this curvature is defined by the shape of the generating curve that the potter chooses. We can think of this curve as marked out on, and cut out from, a sheet of metal. It scrapes against the clay as it turns. It pushes the clay out to the final dimension desired. The outward-pointing radius of curvature of the shard we are examining is identical with the radius of curvature of this generating curve at the point where the scraper worked upon the shard-to-be. We can call this curvature the "generator's own radius of curvature"—or we can call it the swing, a word that conjures up the image of swinging an arc with string and pencil. It contrasts with the "generator-to-axis radius of curvature," or rafter, that we were examining a moment ago.

Measure of "In-the-Surface" Curvature

What is it about the shard—examined by now in such detail—that has any connection with gravity? Its in-the-surface curvature! More briefly, its surface curvature or plaquette curvature, which resembles the curvature of a hiding-face plaquette of Schwarzschild geometry.

Is there some way for us to get from radii of curvature, which we can measure relatively easily, to surface curvature? Yes. By a path of reasoning similar to the reasoning that gives us an expression for the surface curvature of a sphere. As set forth in the box on pages 138 and 139, that reasoning tells us that

$$\text{surface curvature} = -\frac{1}{\text{rafter} \times \text{swing}}$$

Look at two ants, chasing along on the surface of a shard of pottery. They don't know anything about the two principal radii of curvature of that shard or how to measure surface curvature using those radii. In principle, however, they can determine the surface curvature of that shard from the bending of their geodesic courses toward or away from each other, as we saw in Chapter 5. For this determination of curvature, the ants don't have to know anything about the flat or Euclidean three-dimensional space in which their two-dimensional curved shard surface is embedded. For those two ants—let us at least *say*—there *is* no third dimension.

Curvature of Schwarzschild Space Geometry

As for flatland ants traveling on a two-dimensional surface, so for spaceships traveling in the curved three-dimensional space around the Earth, the Sun or a black hole. Neither for ants nor spaceships is there a one-higher-dimension flat space in which they are embedded. For us, however, endeavoring to visualize curved 3-D space, 4-D flat space is a happy enhancement of our powers of visualization. This aggrandizement lets us picture the curved 3-D space of spaceship travel as embedded in an imagined flat 4-D space and endowed—in this flat 4-D space—with three principal radii of curvature.

One of the three principal radii of curvature of 3-D Schwarzschild space imagined to be embedded in a flat 4-D *space* is the swing, the radius of curvature of the generating parabola at a point of interest. The other two radii of curvature are identical in value with the length of the rafter; that is, the distance measured perpendicularly off the given locale of the parabola—through the nonexistent 4-D embedding space—to the axis of rotational symmetry. All three radii of curvature of Schwarzschild space geometry depend on the mass of the center of attraction and on the distance to the point of observation.

It takes a bit of geometry to figure out—from the geometry of the parabola—the dependence of the swing and the rafter on mass and location. Here's the result in cryptic form:

$$\text{swing} = \sqrt{\frac{2 \times (\text{Schwarzschild radial coordinate})^3}{\text{central mass}}}$$

$$\text{rafter} = \sqrt{\frac{(\text{Schwarzschild radial coordinate})^3}{2 \times \text{central mass}}}$$

Two ideas take central place in the derivation of these two radii of curvature. First, the magnitude of the central attracting mass in geometric units is one-fourth the distance from the directrix, or axis of the paraboloid, to the focus of the parabola. Expressed equivalently, the mass is half the distance from the axis to the nearest point on the parabola. Second, all that counts for curvature—in addition to mass—is

FROM RADIUS OF CURVATURE TO SURFACE CURVATURE

Surface curvature displays itself nowhere more clearly, and nowhere more simply, than on the surface of a sphere. Two geodesics, two straightest-possible lines on that curved surface bend toward each other, start out though they do on nearby and perfectly parallel courses. As they advance, the line between them sweeps out on the surface of the sphere an area. Over this area we spread in imagination the responsibility for the bending of the geodesics to determine the curvature of the surface. In other words, we define curvature as the bending associated with a unit area of the surface:

$$\text{surface curvature} = \frac{\text{bending}}{\text{area swept out}}$$

The sphere has this simplicity, that everywhere the curvature is the same. Therefore we need not restrict ourselves to geodesics that run nearby. Let's choose, then, two geodesics that are easy to picture and to analyze. They both start at the equator of the sphere and head straight north—but they run along lines of constant longitude that differ by 90 degrees, one-quarter of the circuit of the equator. Parallel and exactly north-bound though both geodesics start, they nevertheless meet at the North Pole at an angle of 90 degrees, one-quarter of a circle, one-quarter of the full 2π radians of a full circle—that is, at an angle of $\pi/2$ radians. That angle measures the bending that the surface curvature has exerted on the geodesics.

To how much area do we assign responsibility for this bending? To one-quarter of the Northern Hemisphere, or one-eighth of the surface of the entire sphere. And this area of the whole sphere equals 4π times the square of the radius of the sphere. So the responsible area is 4π (radius)^2/8 = $(\pi/2)$ (radius)^2.

Knowing the bending and the area responsible for it, we have all we need to determine the curvature:

$$\begin{pmatrix} \text{surface curvature} \\ \text{of a sphere} \end{pmatrix} = \frac{(\pi/2)}{(\pi/2)\ (\text{radius})^2} = \frac{1}{(\text{radius})^2}.$$

In other words the curvature of a sphere, or of any portion of a sphere—or even of any spherelike portion of a surface of the most irregular shape—is equal to the inverse square of the radius of curvature at that locale. Blow a balloon up to three times its original radius? Then its curvature diminishes—by a factor equal to the square of three—to one-ninth its original amount.

Many a portion of many an object is not spherelike; the two principal radii of curvature are not equal. So conditions are at the belt of a cigar, at the midriff of an American football, at the waist of a cucumber. In such a region, the reasoning that gives us the curvature is a little more complex, but the answer is almost equally simple:

$$\text{curvature} = \frac{1}{\text{one radius of curvature} \times \text{other radius of curvature}}$$

This result applies not only to the positive curvature of cigar, football, and cucumber but also for surfaces of negative curvature, such as a free-standing ring of soap film and the pottery jar we have been examining. For such a surface, the two centers of curvature stand on opposite sides of the surface locale under examination. Therefore, the two radii of curvature are opposite in mathematical sign, and, consequently, the curvature itself comes out to be a negative quantity. The negative sign of this curvature signals that two nearby and nearly parallel geodesics on such a surface bend, not toward, but away from, each other.

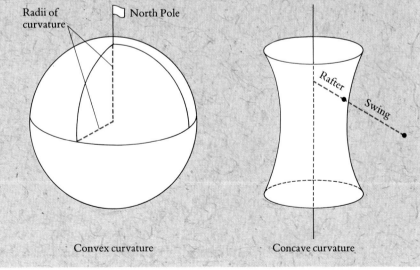

The curvature at any locale on a two-dimensional curved surface can be expressed in terms of its two principal radii of curvature (red dashed lines). Since the centers of curvature lie on opposite sides of a concave surface, the two radii have opposite signs; the resulting negative curvature causes originally parallel geodesics on such a surface gradually to diverge from each other.

Radii of curvature

North Pole

Rafter

Swing

Convex curvature

Concave curvature

location; and all that counts for location is distance from the center of attraction. But distance from the center is crazy physics, crazy terminology, crazy measuring. Travel up and down the parabola as I will, I never get to the directrix, to the axis of rotational symmetry. I never get there as long as I stay on the matter-free Schwarzschild three-dimensional curved-space geometry. However, in the nonexistent flat four-dimensional space geometry I can make, in imagination, a beeline for the directrix and measure straightaway the distance to it. This distance provides one way to conceive of the so-called *Schwarzschild radial coordinate*.

The Schwarzschild radial coordinate is the number-one tool for describing location in curved space geometry. Fortunately there's a way to describe it that requires no travel in never-neverland. Go *around* the axis of symmetry, not in toward it! The circumferential distance so covered, in one circuit, divided by 2π, gives the Schwarzschild radial coordinate of the point in question. It specifies location in terms of real-space distances. Tempted to swear at the cumbersomeness of the ever-recurring term, the S.......... r.......... c.......... ? We can resist that temptation by boiling those three words down to the single letter r. Pushed so far in the direction of algebra, we go the rest of the way and denote the mass of the center of attraction by the letter m.

The pot's two-dimensional surface in a given locale has only two principal radii of curvature: the rafter, pointing in, and the swing, pointing out. However, the three-dimensional curved space outside a spherically symmetric center of attraction when conceived as embedded in a four-dimensional flat space (where the fourth dimension is geometric and has nothing whatsoever to do with time!) has three principal radii of curvature. The length of the rafter provides two of these radii, and they are associated with the two independent directions of around-the-center travel. The swing is the third radius, and it is associated with outbound travel.

How do we get back from the never-neverland of a nonexistent four-dimensional Euclidean embedding space to the real world of curved three-dimensional space? By shifting attention from radius of curvature in that 4-D flat space to surface curvature in that curved 3-D space. Surface curvature, space curvature, plaquette curvature: that's the curvature that matters for gravity. That's the curvature where the grip of gravity lies, shows, and works. Gravity doesn't act at a point. It acts in a locale, in a locale where spacetime geometry is curved. No free-float pebble, no free-float neutral particle, no free-float elevator, Einstein taught us, feels gravity. Two nearby and nearly like-directed free-float test masses, however, reveal gravity—reveal it by the relative bending of their worldlines, their geodesics, their free-float tracks through spacetime. That bending evidences spacetime curvature. That bending, expressed on a per-unit-area basis, does more. It measures curvature. Such in-surface curvature is the hallmark of gravity.

As we've done before, let's imagine a locale in curved space as a small cube bounded by six faces or plaquettes (see page 129). The curvature of spacetime in a given locale is defined by the two principal in-surface curvatures of its bounding plaquettes. Let's orient our little cube enclosing any given locale in Schwarzschild space geometry so that two of its plaquettes face the attracting central mass head-on; these are the exposed plaquettes. The other four plaquettes, the hiding ones, are oriented edgewise to the center of attraction. The surface curvature on an exposed plaquette and hiding plaquette are given by the following expressions:

$$\binom{\text{surface curvature on}}{\text{an exposed plaquette}} = \frac{1}{(\text{rafter})^2}$$

$$= \frac{1}{\left[(\text{Schwarzschild radial coordinate})^{3/2}/(\text{twice central mass})^{1/2}\right]^2}$$

$$= \frac{1}{(\text{Schwarzschild radial coordinate})^3/\text{twice central mass}}$$

$$= \frac{\text{twice central mass}}{(\text{Schwarzschild radial coordinate})^3}$$

$$= \frac{2m}{r^3}$$

$$\binom{\text{surface curvature on}}{\text{a hiding plaquette}} = -\frac{1}{\text{rafter} \times \text{swing}}$$

$$= -\frac{(\text{twice central mass})^{1/2}}{(\text{Schwarzschild radial coordinate})^{3/2}}$$

$$\times \frac{(\text{half central mass})^{1/2}}{(\text{Schwarzschild radial coordinate})^{3/2}}$$

$$= -\frac{\text{mass}}{(\text{Schwarzschild radial coordinate})^3}$$

$$= -\frac{m}{r^3}$$

Any little mass-free cube at any locale in the space around a central attracting mass can be oriented so two of its plaquettes are fully exposed and the other four maximally hide. For each exposed plaquette, there are two hiding plaquettes. For the surface curvatures to compensate, which they must, the surface curvature in each hiding plaquette is half what it is in an exposed plaquette.

The grip of spacetime on momentum-and-energy, acting in a region where there *is* no momenergy to act on, acts on spacetime itself. Curvatures have to compensate, and in Schwarzschild geometry, by virtue of the parabolic-generator construction, curvatures *do* compensate. That's the story of Schwarzschild space geometry.

Apollonius of Perga's Discovery of the Key to Black-Hole Geometry

How does black-hole 3-geometry differ from soap-film 2-geometry? A bit of soap film can attain equilibrium only when its radii of curvature stand out perpendicular from it with opposite directions and equal magnitudes. A spherically symmetric empty-space 3-geometry can remain as it is only when the tugs of curvature upon it compensate; that is, only when the two relevant radii of curvature stand out perpendicular on opposite sides of the generating curve, one *twice* as great as the other.

It is amazing that the law of the space outside Earth, Sun or black hole differs from the law of the soap film by only this simple factor of two! But the "two" is essential! One radius of curvature has to be double the other. That circumstance is justification enough to call our finding *the law of double curvature*.

I had never grasped this double-curvature condition until one day in Austin, Texas, led on step by step by questions from the bright-eyed students before me, I eventually found myself writing on the blackboard something quite new to me! More than one student had managed by the next day to work out by means of calculus that one curve, and one curve only, meets this condition: the parabola. But surely, I thought, history did not give us first the calculus and then the properties of the parabola, but first the properties of the parabola and then the calculus. Surely someone somewhere at sometime knew that the parabola of old satisfies our modern-day demand for double curvature. I asked Marshall Clagett, distinguished historian of science, and he advised me to look into Book V of the *Conics* of Apollonius of Perga. There I found, among the sages of old, "the glory that was Greece," a new perspective on the parabola, generator of the most important curved space geometry in all of physics: Schwarzschild geometry, the geometry around a spherically symmetric center of attraction.

Apollonius came from Perga on the coast of Asia Minor, not far from Ephesus. There, a century earlier, Alexander and his forces had turned inland to defeat Darius and change the history of the world. Apollonius lived most of his creative life in Alexandria, around 247 to 205 B.C. He studied, among other things, the three kinds of sloping slices that can be cut through a cone. To these he gave the names that are still today the standard terms: *parabola,* for the section that is parallel to the opposite edge of the cone; *ellipse,* for the section whose slope is less than that of the opposite edge; and *hyperbola,* for the section whose slope is greater than that of the opposite edge (see illustration on page 144). On the properties of these curves, Apollonius wrote an eight-book treatise, *Conics,* which earned him the title of the Great Geometer. This treatise had a fame in the ancient world comparable to Euclid's thirteen-volume *Elements.*

Proposition 99 in Book V of the Arabic translation of *Conics* gives the proof that at any point on a parabola the radius of curvature in the plane of the curve is twice as great as the distance, measured perpendicular to the curve, from that point to the directrix. In brief, Proposition 99 states that the parabola exhibits double curvature. Edmund Halley

The wall and two obelisks visible in this picture of Alexandria from around 1800 had survived from ancient times. They were daily seen—as was, too, the tomb where the embalmed body of Alexander himself lay—by Apollonius on his way to the school of Euclid and the Library of Alexandria about 247 to 205 B.C. One of the two obelisks stands today on the Embankment in London; the other in Central Park, New York City.

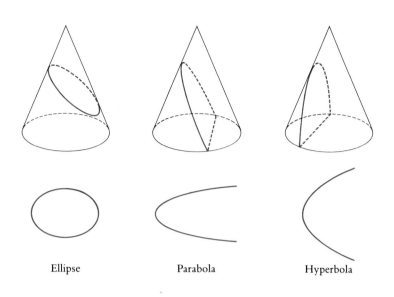

Ellipse Parabola Hyperbola

(1656–1742), famous for his own contributions to science, for his encouragement and help to Newton in the publication of the *Principia,* and for his demonstration that comets are ordinary members of the solar system, moving in elongated elliptical orbits, and returning periodically, learnt Arabic expressly so that he could translate the Arabic version of Apollonius into Latin, accessible to every learned person of his day. In the renumbering of Halley, Proposition 99 of the Arabic appears as Propositions 51 and 52 of Book V.

My study of Proposition 99 has depended neither on the Arabic translation nor Halley's Latin translation. But even in today's English edition of *Conics,* edited by T. L. Heath, and in C. B. Boyer's elucidation of the proposition in his *History of Mathematics,* I have found it difficult to see that Apollonius does in fact show that a parabola has double curvature. Nowhere does Apollonius use the concepts of directrix, focus, curvature, or radius of curvature, which seemed to me so essential in demonstrating this property of the parabola. Nevertheless, with time and help—not least from Debra Saxton, an Austin high-school student in love with telescopes and telescopic instrumentation—I learned that it was all there. Apollonius *does* prove, really prove—without ever explicitly saying so—that the radius of curvature of a parabola at any point is twice the distance from that point to the directrix.

My non-calculus, geometric way of demonstrating the double curvature of the parabola uses the old-fashioned compass and ruler approach. We start by constructing a parabola for which the distance from directrix to focus is 4 × 0.444 centimeters = 1.676 centimeters. This dimension is just right for representing the geometry that we would get if we could completely collapse the Earth to a black hole. At any point *P* upon that parabola (see illustration on page 131), we draw in the line (not shown) tangent to the parabola. From that tangent, we then construct the perpendicular and extend it until it crosses the directrix at D. The distance PD marks one radius of curvature of the black-hole space geometry generated by cranking the parabola about the directrix. Our question is clear and simple. Does the other radius of curvature, extending away from the directrix, have twice this length?

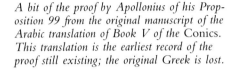

A bit of the proof by Apollonius of his Proposition 99 from the original manuscript of the Arabic translation of Book V of the Conics. *This translation is the earliest record of the proof still existing; the original Greek is lost.*

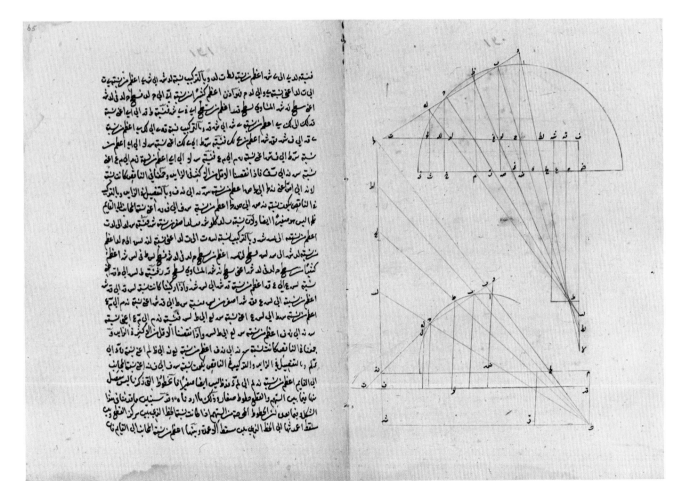

Tracing out a parabola by unzipping a zipper. A thumbtack holds one tab of the zipper fixed at the focus, F, of parabola. The other tab, gliding without friction on the directrix, D, keeps level with the slider of the zipper. That slider, via an attached pen, traces out the parabola. Each point of it lies equidistant from focus and directrix.

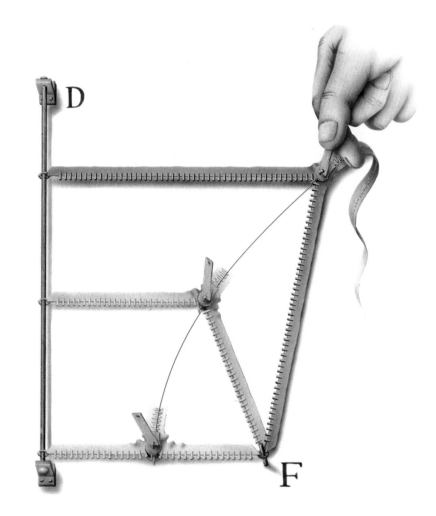

We get out our trusted compass. We plunge the metal-tipped point into the downward extension of the perpendicular at some trial point. We adjust the other arm of the compass so that the sharpened pencil point hits the point of crossing of perpendicular and parabola. We swing the pencil point to make a great circular arc. Oops! The resulting circle falls entirely within the parabola. So we try again with another, more distant point for the center of our arc. And again. Eventually we end up with a pinprick at just the right point on the perpendicular. The circular arc centered there "osculates" or kisses the parabola over the longest possible arc. If we place the centerpoint any farther away from the directrix, then the length of the coinciding arcs will begin to decrease. Thus we have found the center of curvature (C) of the parabola at our randomly chosen point P. Having found the center of curvature, it is a simple matter to show by direct measurement that the in-the-plane radius of curvature (PC) is twice as great as the cranking radius of curvature (PD) about the directrix. And if it's true for one randomly chosen point, then we'll assume it's true for all points. Now we can see for ourselves that the parabola does indeed have double curvature, the property that inextricably weds this particular curve to Schwarzschild, or black-hole geometry.

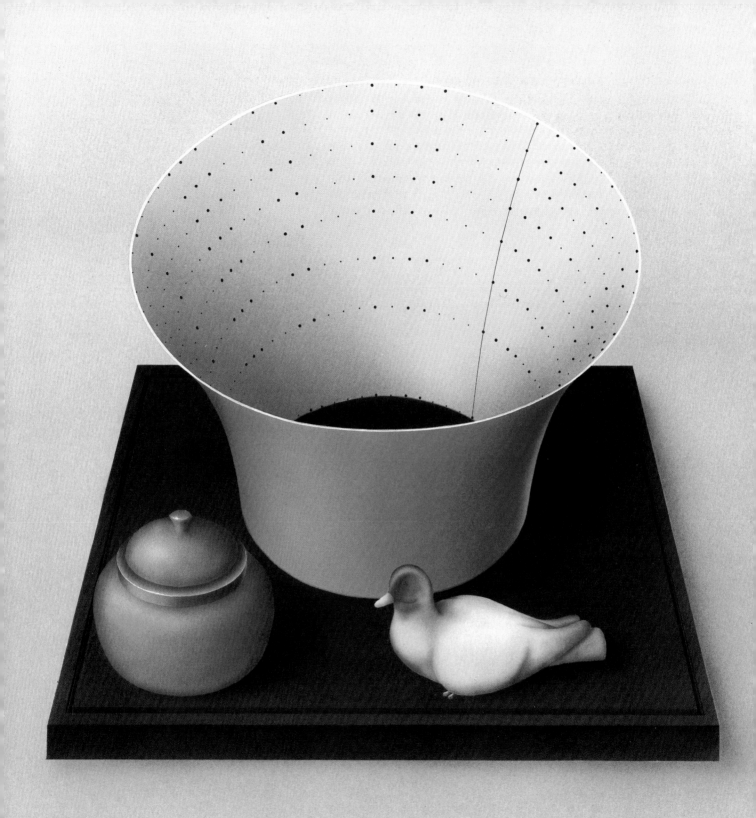

9 Picturing Space and Spacetime around a Center of Mass

Gravity, you have led us far
From the boundary of a boundary is zero
To momenergy as moment of rotation.
Now, at the end of the trail,
We capture your greatest prize,
The spacetime geometry
Around a center of attraction,
Guide to the workings of Earth and Sun,
Black hole and cosmos.

The bottomless porcelain paraboloid symbolizes the curved space around Earth, star, or black hole. The red dots mark locations on the curved surface. Measuring the distance between the dots reveals the curvature of the surface.

In one of my dreams, my granddaughter Elizabeth and I are admiring on the table before us the pottery paraboloid, with its beautiful smooth green surface. It symbolizes, we know, the curved space geometry around Earth, star, or black hole. What must it be like to navigate through all the emptiness of space? How is our spaceship pilot to know where we are? Or know how long we must travel to get from A to B? No marker of place, no signpost of distance can we imagine in curved empty space, and none do we see anywhere on that glassy greenness.

Elizabeth suddenly catches sight of a dark ring rising slowly up out of the throat of the paraboloid. A closer look discloses that the ring is not a ring. It is three-and-sixty humming soldier ants, marching upward side by side, in the formation of a phalanx that bends around to make a circle. From time to time the humming stops an instant as all 63 ants arch up each to deposit a red dot of marker fluid on the green surface. Fascinated, we watch to the end, when the ants tumble off the rim of the jar. We dispose of them and turn back to admire the regularity of the pattern they have left.

"Those red location markers on that two-dimensional curved space," I smile to Elizabeth, "what better symbol could we want for markers of location in curved three-dimensional space?"

She has an eye for pattern and smiles, too, but clearly she spots a feature of the pattern that escapes me. Fetching a ruler, she shows me. "The 63 dots around the innermost circle have a separation, one from another, of close to 0.9 millimeter. In the next circle the separation from dot to dot is about 1.0 millimeter; in the next, 1.1 millimeters; in the one after that, 1.2 millimeters; and so it goes. You were talking a few minutes ago about the need for a way to mark position in space, and the red dots illustrate what you mean, but you also spoke of needing a way to measure distance, of needing what I think your friends call a *metric*."

"Yes, metric," I reply, "a formula that tells us the distance between point and nearby point. The collection of distances thus supplied reveals the shape of space or spacetime geometry even better than chest, waist, and hip measurements reveal the figure of a human. No mathematical device is known that better describes one curved spacetime geometry and distinguishes it from another."

"Wonder of wonders," Elizabeth rejoins, "haven't the ants themselves already supplied us with the beginnings of such a device? These soldier ants are far ahead of the West Point military cadets who, in line, space themselves off, finger tips against finger tips. These ants didn't have to push apart. Each traveled a line as straight as it possibly could up out of the throat of that green jar. The very curvature of the surface bit by bit automatically enlarged the circumference of the circle and—with it—the spacing between ant and ant. By some miraculous sense each ant gauged his distance from his partner on the left—and from his partner on the right. Every time that distance increased another tenth of a millimeter, each ant made his mark. Those ants give us location markers, and sideways distances, too. But as for distance of progress up out of the throat of the jar, no ant seemed to care about *that!* The record they've left us is not complete enough to reveal the distance from each point to all its near neighbors. A start on a metric, but only a start!"

The Schwarzschild Space Metric

The dream fades. Before all memory of it goes, I make a picture to remind me of the pattern of those dots, and the faint red line made by the ant whose ink depositor was evidently leaking. In the picture I suddenly see unfold the whole story of the Schwarzschild space metric of the geometry around a spherically symmetric center of attraction. First, *coordinates,* radial and angular, identify a point, name it, single it out from all other points. Second, *small changes* in those coordinates identify a second point in the close vicinity of that first point. Third, the *metric formula* itself figures the distance—which ordinarily runs slantwise—from the one point to the other. That metric formula is a generalization of Pythagoras' formula for the length of the hypotenuse of a right triangle: $(\text{side } 1)^2 + (\text{side } 2)^2 = (\text{hypotenuse})^2$, or in algebraic notation, $a^2 + b^2 = c^2$.

All of the red dots in any one circle have in common the same value of the "reduced circumference" or "Schwarzschild radial coordinate," commonly abbreviated to a single letter, r. I put my finger on that circle of 63 dots between which Elizabeth, in my dream, measured a separation close to 1.0 millimeter. This circle has a circumference of very nearly 63×1.0 millimeter = 63 millimeters. By definition, the reduced circumference is the actual circumference divided by the factor $2\pi = 6.2832\ldots$ (that is, divided by about 6.3), and so the reduced circumference of the circle is close to 10 millimeters. This number gives us the Schwarzschild r-coordinate of each and every point on the circle. From these points, the points on the next smaller circle are distinguished by an r-coordinate of 9 millimeters; those on the next larger circle, by an r-coordinate of 11 millimeters; those on the following circle, by an r-coordinate of 12 millimeters; and so on.

In the forming of the paraboloid of revolution, the clay turns around a central axis. Were that axis reinstalled, one interpretation of the r-coordinate would stand out for view: the straight-line distance from red dot to axis. To the ants traveling in the two-dimensional curved space of

Pythagoras' formula, generalized to a limited locale of a curved surface, gives the metric for figuring out the distance from point to point.

The r-coordinate and the angle coordinate identify the black-encircled red point on the green paraboloid.

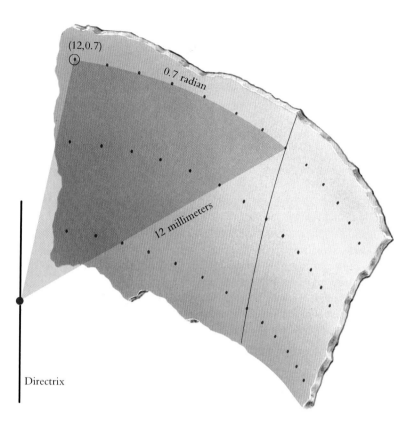

(12,0.7)

0.7 radian

12 millimeters

Directrix

the jar's surface, however—as to us, were we traveling through the three-dimensional curved space around a black hole or star—there is no axis "out there," no extra space dimension available to reach out through to an axis. Were we to speak of "*r*" as point-to-axis distance, we would make a statement with no testable meaning. In contrast, circumference is a quantity we *can* determine by measurements within the curved space, and the definition of *r* as circumference divided by 2π is sharp and clear. This reduced circumference supplies the first of the two coordinates we need to distinguish point from point.

The second coordinate of a point is angle. Angle distinguishes points in the same circle, one from another. The black-encircled red dot on the green surface is seventh around the circle. How much is its angle coordinate? The whole angle around any circle being 2π, or close to 6.3 radians, and there being 63 points around the circle, each point separates itself from the one before it by very nearly one-tenth of a radian. So the angle coordinate of the selected point is 0.7 radian. It lies on the circle of reduced circumference $r = 12$ millimeters. Therefore, the complete name of this point in the (r, angle) identification scheme is (12, 0.7). The next point around the circle is (12, 0.8). Not one red dot stands unidentified on the whole green surface.

When we turn our attention from the green 2-D geometry to the 3-D curved space outside a black hole (or any other essentially spherically symmetric center of attraction) and consider all the points that have the same Schwarzschild r-coordinate, we find that those points constitute, not a circle, but a sphere. To distinguish those points requires, not one angle, but two, for example, longitude and latitude. However, as we orbit in free-float motion around that black hole or some star or the Earth itself, we automatically stay in whatever the plane of motion we started in. We need only one angle to describe our progress in that plane, not two. For this reason, and for the sake of simplicity, it's reasonable to limit ourselves in this book to a single angle—plus the r-coordinate—in exploring how to measure distance from point to nearby point in this 3-D curved space.

"Nearby" is no trivial idea. To state it properly leads us to the second feature of metric, the concept of a change in coordinates from point to nearby point, a change so small that it counts as "differential," a concept traditionally abbreviated dr for differential of r-coordinate and $d(angle)$ for differential of angle coordinate.

Why do we consider only nearby pairs of points? Why do we require the associated coordinate differences to be small, differentially small? To see why, we don't have to look at curved space; it's enough to see how well the angle and r-coordinates determine distance between two points, A and B, that happen to lie on the same circle on a flat piece of paper. When the points are close enough to entitle us to use the concept of differential of distance, $d(distance)$, and differential of coordinate, $d(angle)$, then it is correct to speak of the elementary distance between those two points as being

$$d(distance) = r \times d(angle)$$

It is wrong in principle to apply this formula to figure the distance between A and B when the angle is finite. For example, when those two points lie directly opposite each other on the circle, separated by a full half-circle having an angle of π radians, then the correct figure for the distance between them is a full diameter, or $2r$. Yet the formula $r \times$ angle would give $\pi \times r$, or $3.14 \times r$, an error of more than 50 percent. When, however, the angle between A and B is as small as it is between the dots made by two partner ants, 0.1 radian, then the formula gives $0.1 \times r$ for the arc distance, whereas the correct figure for the straight-line distance is $0.099996 \times r$. The error itself is small. More important, the percentage error is small—only four-thousandths of one percent.

To be able to attain such precision is the payoff for dealing only with distances between nearby points. We can then add these distances to get the distance between two more widely separated points. Percentage error is not increased by the adding process. And percentage error, though not zero, can be made as small as we decide to make it by subdividing the total track of interest into sufficiently many sufficiently short pieces. That is the wisdom hidden in those innocent-sounding terms: nearby points, small coordinate change, coordinate differential, and distance differential.

The essence of these concepts we owe to Newton and Leibniz. Independently, and well before 1700, they invented what we today call the calculus. It took at least a hundred more years to state every little bit of the reasoning in objection-free and battle-tested language. Deal always with "vanishingly small quantities" to get precise results? One famous critic of Newton, Bishop George Berkeley (1685–1753)—original thinker about the nature of gravity and source of the name of the city associated with one great campus of the University of California—railed against the calculus as dealing with "the ghosts of departed quantities." To deflect such criticisms it helped to switch—as all the world since has done—to the notation of Leibniz and the symbol, dr, for "differential of the radial coordinate, r." To expound the sophisticated sense in which the notation "dr" has a thoroughly justified meaning is no trivial matter.

We skip the sophistication and adopt the notation. For quick insight, we'll cavalierly call Leibniz's differential dr "the small change in the radial coordinate r"—and, more generally, let Leibniz's symbol "d" itself stand for the handy phrase, "small change in . . . ." We can leave the elucidation of that word *small* to this or that textbook of calculus!

When we turn from small change in angle to small change in r-coordinate, we get a new payoff from adhering to the demand that the change be *small*. Apparently small, but not adequately small, was the change in r-coordinate of the ink-dripping ant as it moved from the circle of $r = 9$ millimeters to the circle of $r = 10$ millimeters, and from there to the circle of $r = 11$ millimeters. The distance actually covered in each of these two steps is plainly greater than one millimeter, and greater, moreover, in the first step than in the second. In brief, the green paraboloid, in the vicinity of the throat, rises upward much more than it spreads out-

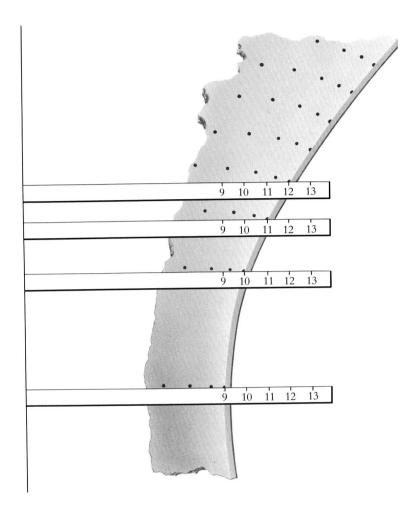

The proper distance from red dot to red dot is traced out by following the curved green surface. The ruler indicates the increase in the r-coordinate from dot to dot. Clearly the proper distance is longer than the increase in the r-coordinate.

ward. This feature of the parabola lends itself to capture in a simple formula when—but only when—we limit attention to a *small* change in *r*-coordinate:

$$d(distance) = \frac{dr}{\sqrt{1 - 2m/r}}$$

Here *2m,* the *r*-coordinate of the throat itself, is twice the mass of the center of attraction, expressed in geometric units. For the Earth itself, the

ONE SMALL STEP ON THE WAY
UP OUT OF THE THROAT

The quantity $d(distance)$, gives us the distance traversed in one small step in the climb up out of the throat of the green porcelain paraboloid—or outward through the empty-space Schwarzschild geometry that this ceramic models. Illustrated in the diagram as the distance between the two black dots in the magnified view, $d(distance)$ is larger than the increase, dr, in the Schwarzschild r-coordinate, and it is larger in the ratio of 1 to $\sqrt{1 - 2m/r}$. The origin of this famous factor we uncover here, not by the traditional higher-level methods of calculus-plus-differential-geometry, but by such elementary reasoning as Apollonius already was accustomed to employ in 220 B.C., based on nothing more than the comparison of two similar triangles. Those triangles are identified here by their common color, pink.

The short side of the pink triangle in the magnified view is the small increment, dr, in the distance, r, from the directrix to two arbitrarily chosen, yet very close, points on the parabola. The quantity r is what we variously call the reduced circumference or the Schwarzschild r-coordinate. The little triangle's long side or hypotenuse is the small distance, $d(distance)$, of real-world outward travel required to bring about that increase in the r-coordinate. We want to find the ratio $dr/d(distance)$. The two points in the magnified view are so close that in the bottom diagram they appear as one point, shown underneath the small magnifying glass. This point is the vertex of another pink triangle. If we can show that the two pink triangles are similar, then the length of two of the sides of the larger triangle will give us the ratio we search. In the magnified view, the two blue lines that run down to the focus of the parabola have—by the very definition of a parabola—the same length as the two brown lines that run straight left to the directrix. Because the length $d(distance)$ is small in the sense of calculus, we can count the two blue lines as parallel. This granted, we are assured that the hypotenuse of the little triangle bisects the angle between the brown $r + dr$ line and the blue $r + dr$ line. Switching to the larger diagram, we prolong the straight-line bisector defined by this hypotenuse (dashed pink line) downward to the lower left to the point where it meets at a right angle, and bisects, the dotted pink line that runs from the focus to the point where the cranking radius, r, meets the directrix. This bisector constitutes the short side of the larger pink triangle. The two pink triangles are similar because they have two equal angles: the angle formed by the brown and dashed pink lines and the two right angles. The ratio of the length of the brown line and the dashed pink line in the larger triangle gives directly the stated ratio between dr and the actual distance, $d(distance)$, covered in the outward step. We know the length of the brown line is r. But what is the length of the dashed pink line? (Continued on next page.)

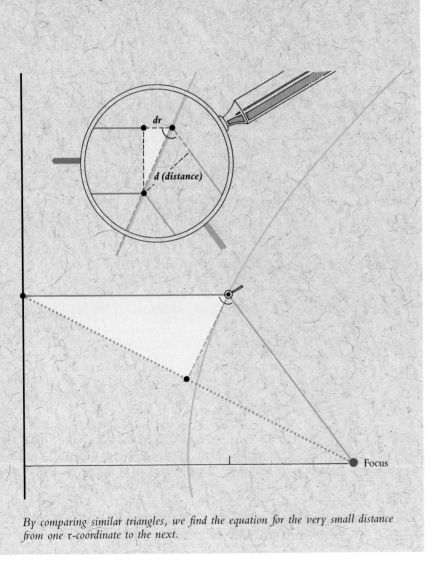

By comparing similar triangles, we find the equation for the very small distance from one r-coordinate to the next.

The projection (or shadow cast) by the brown and blue lines onto the horizontal axis have to add up in length to the directrix-to-focus distance, which is four time the mass of the center of attraction, or $4m$. Therefore, the horizontal projection of the blue line by itself is $4m - r$. The square of this horizontal projection is $16m^2 - 8mr + r^2$. The square of the length of the blue line itself is r^2. The difference of the previous two squares is $8mr - 16m^2$, which gives the square of the vertical projection of the blue line. The dotted pink line has the same vertical projection. Its horizontal projection is $4m$, which has for square $16m^2$. The sum of the two last mentioned squares gives the square of the length of the dotted pink line, $8mr$. The lower side of the pink triangle has half that length. Therefore, its squared length is one-fourth as great, or $2mr$. The hypotenuse of the pink triangle has length r, and squared length r^2. Taking the difference of the last two squares, we get the squared length of the short side of the pink triangle, $r^2 - 2mr$, and from it that length itself, $(r^2 - 2mr)^{1/2}$.

The hypotenuse, r, stands to this short side of the big pink triangle in the ratio $1/(1 - 2m/r)^{1/2}$—and therefore this also is the ratio of hypotenuse, $d(distance)$, to short side, dr, of the little pink triangle. In this ratio

$$d(distance)/dr = \frac{1}{(1 - 2m/r)^{1/2}}$$

there stands before our eyes the epitome of the curved-space geometry around any spherically symmetric center of gravity—and derived without appeal to calculus or differential geometry.

quantity $2m$ is twice 4.4 millimeters, or 8.8 millimeters. That's the radius that the potter chose for our green vase. Its two-dimensional geometry symbolizes the character that the three-dimensional space geometry around the Earth would have if our planet were collapsed into a black hole, so that we could study its geometry all the way in to the throat! The space geometry outside our present-day uncollapsed Earth corresponds, naturally enough, only to the outer and less curved portions of the green paraboloid.

Now for Pythagoras's rule, $a^2 + b^2 = c^2$, and the third and final part of the idea of metric! The separation between point and nearby point in the motion of an object will normally run neither purely in the direction of increasing r-value, nor purely in the direction of increasing angle, but partake of both displacements, and run at a diagonal. To discover the length of that diagonal—or, quicker and easier, the square of that length—Pythagoras instructs us to square the two displacements individually and add:

$$d(distance)^2 = \frac{dr^2}{(1 - 2m/r)} + r^2 d(angle)^2$$

This is the space part of the famous Schwarzschild metric, squeezed into the compact algebraic notation that by now has become standard. Legalistically it is wrong. On the left should appear $[d(distance)]^2$, that is, the square of a small distance, not $d(distance)^2$. If we were creduluous enough to take it seriously, $d(distance)^2$ would give us the crazy idea of the small change in the square of a distance—distance from Lord knows what faraway starting point. How did this boo-boo come about? Pure laziness. People got tired of writing down those square brackets, left them out, whispered a warning to their friends to write them back in—mentally, at least—when putting Schwarzschild's metric formula to use, and by now we're all in on the little secret. The same with the terms dr^2, which rightfully should read $[dr]^2$, and $d(angle)^2$, which means $[d(angle)]^2$. The last quantity in the formula, $r^2 d(angle)^2$, is to be understood as $[r \times d(angle)]^2$, or exactly what Pythagoras calls for, the square of the circumferential separation between the two nearby points. To this second term we add the first term, the square of the radial separation between the two points. The addition of both terms indeed gives the square of the distance between them.

It is no small achievement to have won a knowledge of the distance between every point and every other nearby point throughout the whole paraboloid-like space. Those points and those distances are not a commentary on space geometry. They *are* the geometry.

Space is like a Buckminster Fuller dome. Give the erector the junctions and the connector beams. Let each beam have its predetermined length. Let each end be marked to indicate the junction point to which it is to be welded. These assignments made, the shape of the resulting dome is thereby fixed, even if that shape is something so preposterous as an enlarged version of a double-humped camel back. Metric is the encapsulation of points plus distances, and metric defines geometry.

Schwarzschild Space Geometry in Retrospect

By the pattern of point-to-point distances that it enfolds, the Schwarzschild space metric defines the geometry of our Apollonius-inspired superparaboloid. Moreover, that's the only geometry there is that satisfies the twin demands we make. First, the geometry must be spherically symmetric, as befits space around a spherically symmetric center of attraction. Second, in every locale where space is empty the geometry must be so curved as to annul in that domain the moment of rotation—because where space is empty the content of momenergy is zero.

Textbooks on Einstein's geometric theory of gravity employ the mathematical machinery of the calculus and differential geometry to show that the Schwarzschild space metric does indeed satisfy the demand for zero moment of rotation. We've already established the same point with more insight and without having to break into that mathematical territory. It was enough for us in Chapter 8 to draw pictures and—helped by Apollonius of Perga—use considerations of everyday geometry. That's how we showed that the curvature of any chosen plaquette of 2-D ceramic surface, expressed in radians per square meter, is −(mass)/(distance from the cranking axis)^3 or, still briefer, curvature $= -m/r^3$. In the curved empty space around a center of attraction, we confronted at any chosen locale, not one plaquette but three distinct plaquettes, each perpendicular to the other two and each with its own measure of curvature. On each of the two that stand edgewise to the center of attraction the curvature is as just written. On the one that faces the center full-on, the curvature is twice as great in magnitude and positive in sign.

The sum of all three curvatures is required to be, and is, zero:

$$ -\frac{m}{r^3} - \frac{m}{r^3} + \frac{2m}{r^3} = 0 $$

A zero moment of rotation attaches to the little *spacelike* cube defined by the three mutually perpendicular plaquettes. It contains no mass, no momentum, and no energy for spacetime to grab and grip.

From Space Geometry to Spacetime Geometry

Metric must, can, and does generalize itself from curved space to curved spacetime. It is spacetime, not space, that tells rock, spaceship, and planet how to move. To describe and understand properly the spacetime that surrounds any great spherically symmetric center of mass, we must add a new term to our metric specifying time.

The full Schwarzschild metric formula for the geometry of spacetime reveals itself as

$$d(interval)^2 = (1 - 2m/r)dt^2 - \frac{dr^2}{(1 - 2m/r)} - r^2 d(angle)^2$$

This formula reveals something more important for motion than time alone and something more important, too, than the distance from point to nearby point. It reports the spacetime interval racked up by a real or imagined test mass that passes from arbitrary event A to an equally arbitrarily chosen *nearby* event B. And to report the interval is to disclose the advance of wrist-watch time registered and the aging experienced by the particle.

Such tiny intervals, such bits of aging, can be added up along the entire course of a conceivable history of a test mass advancing through spacetime. The sum of the tiny intervals reveals to nature the total aging that the particle incurs while running through that history. The magnitude of total aging decides which of all conceivable motions the test mass will follow. Nature magically compares the aging caused by each conceivable motion with the aging caused by all the others and selects the actual motion out of a myriad of possibilities. In brief, spacetime exerts its grip on mass by paying attention to spacetime interval—and interval reveals itself in metric.

Reaching beyond Space Curvature to Spacetime Curvature

Where does Schwarzschild's wonderfully compact formula for spacetime metric come from? From considerations of emptiness, of zero content of momenergy, of zero moment of rotation, like those a few pages back that led to the Schwarzschild space metric.

It is not an adequate description of the region outside the center of attraction to say that there's no momenergy presently there. We want to recognize also that there's no momenergy *flowing* into that vicinity. Otherwise stated, a center-facing plaquette, followed for a small interval of time, sweeps out a three-dimensional block of spacetime. This block has one time dimension and two space dimensions. It too must manifest a zero content of momenergy, and therefore a zero moment of rotation. The curvatures on three perpendicular plaquettes, selected from the faces

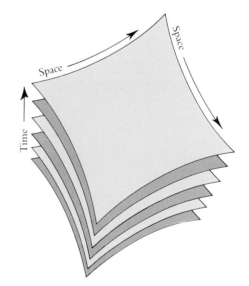

During a small interval of time, a center-facing plaquette sweeps out a three-dimensional block of spacetime.

of that spacetime cube, must add to zero. On one of those plaquettes, the center-facing one, space-space in character, we already know the curvature, $2m/r^3$. A second plaquette has one spacelike edge, but its second edge runs in the direction of increasing time. Likewise for the third plaquette. By reason of spherical symmetry, curvature on this third plaquette has to be identical to curvature on the second plaquette. In any region empty of momenergy, the sum of the curvatures on those two time-space plaquettes has to compensate the curvature on the center-facing space-space plaquette. On each time-space plaquette, therefore, the curvature must be $-m/r^3$. It turns out that this requirement fixes the ratio between lapse of proper time, or differential interval, or increment of aging, on the one hand, and lapse of global time coordinate t:

$$d(interval) = (1 - 2m/r)^{1/2} \, dt$$

We have just discovered that it is not enough for space to be curved. Spacetime itself has to be curved. Only so can we guarantee zero content of momenergy in spacetime blocks of every variety outside the attracting mass. Moreover, it's not the curvature in a space-space plaquette, but the curvature in a time-space plaquette that dominates in determining the rate of separation, or approach, of two nearby objects so slow compared to light as spaceships and planets. So we learned from the analysis of boomeranging in Chapter 4. And it's the curvature in a time-space plaquette that we have just now figured out, $-m/r^3$. It could never exist were there not the factor $(1 - 2m/r)^{1/2}$ that connects proper time and global time. This factor is analogous to, but more important in its consequences than, the factor $(1 - 2m/r)^{-1/2}$ between true outward distance and r-coordinate increase, dr.

In summary, spacetime outside the attracting mass shows everywhere zero moment of rotation in every little cube, whether it's a space-space-space cube or a time-space-space cube. No mass there, no momentum, no energy, for spacetime to grab and grip. Spacetime at one place has to fasten its grip on nearby spacetime alone. Grip and countergrip have to balance out, sum to zero, cancel, and they do. That feature and spherical symmetry distinguish Schwarzschild spacetime in simplicity and importance from every other manifestation of the grip of gravity.

Our paraboloid of revolution supplies a description of the space part of the Schwarzschild geometry that would have delighted Apollonius of Perga with its pictorial quality. The time part of the Schwarzschild metric is even more decisive in its influence on planetary motion, yet spacetime geometry is no tame object that admits of any tame spacelike picture. Its essence is life and movement, the movement of spaceships, rockets, and planets. In those motions we have the most vivid of all evidences of what spacetime geometry is and does. To turn now to planetary motion is the best possible way to picture the grip of spacetime on mass.

10 Stones in Flight and Planets in Orbit

Oh Mars, red though you are,

And symbol of war

Though you seem to some,

For us you are witness

To that synchronic double motion

Of every mass

Attracted by another

Which exulting Kepler first saw in you.

In and out you go

In such a rhythm

That your path

Is an ellipse.

Round and round you also go

At a speed that

Rises and falls so perfectly

That every hour,

As seen from the Sun,

You sweep out

The same area in space.

Reveal to us

The guiding principle

Of your motion,

Energy hill

Master of fall of stone,

Tick of clock,

Orbit of comet

And crash of mass

Into black hole.

Jupiter with Io, Europa, and Callisto, three of its four Galileo-discovered moons. Galileo proclaimed to all that in Jupiter and its moons we see in miniature the working of the solar system.

Despite all we've been learning about the curved spacetime geometry around a spherically symmetric center of attraction, a little round stone lying on the pavement sows doubt. Its wink and leer seem to say, "You much too learned people! You take away Newton's simple gravity. In its place you give us back curved empty spacetime. And at the end, as summit of your supposed learning, you produce a single formula whose only payoff, that I can see, is a supposed aging. What does that have to do with my rise and fall when you toss me?"

We look at each other with interest and then back at the stone. I agree, "Fair enough! Let me turn your commentary into a question that I have a chance to answer. Let me throw you straight up using such a force that you return to me in exactly 2 seconds. During that time you will climb 4.9 meters. First you are in my hand, now, event A, and you're back here 2 seconds from now, event B, and in between you're in a condition of free float. Then why do you have to climb into the air precisely 4.9 meters?"

"Preposterous," the stone winks back. "If you can explain that feature by the principle of maximum aging, you can explain anything about gravity."

We accept the wonderful challenge to demonstrate that on any other obvious course between A here, now, and B, here, 2 seconds from now, the stone's aging would be less than the one it actually experiences! A free-float path through spacetime racks up on a comoving watch an amount of proper time greater than that taken by any modestly different alternative motion that connects the same initial point in spacetime with the same final point. Nature chooses the history, the track through spacetime, that is the straightest, that accumulates the most aging. The stone's question puts us onto the best possible way to see how what we call gravity really arises. Aging depends on place—even for a person who stays always in the same place.

The Principal of Maximal Aging

To discuss time properly we picture a place so far from all disturbing mass—and yet so immobile relative to our center of attraction, the Earth—that we can use a clock located out there as a well-defined standard of reference for the measurement of time. It beams to us a signal to mark the beginning of a one-trillion, or 10^{12}, tick run and a second signal to report the end of the run. Between the arrival of those two signals, how many ticks, or how much aging, will an identical clock here on Earth rack up? The same? or more? or less? To answer the question we don't have to know how far away the distant clock is or how long the start-of-run signal took to travel from the clock to us. Because the end-of-run signal takes the same time, the signaling time cancels out in a comparison between two clocks. The faraway clock registers global time. Our Earth-supported clock reports aging, the aging undergone right here in a specified run of global time. The aging racked up on the Earth-bound clock between the two signals will be less than the aging, or global time, racked up by the faraway clock between the same two signals. According to the aging formula encapsulated in the Schwarzschild metric,

$$\begin{pmatrix} \text{aging accumulated} \\ \text{by Earth-bound clock} \end{pmatrix}$$

$$= \sqrt{1 - \frac{\text{twice mass of Earth}}{r\text{-value of clock location}}} \times \begin{pmatrix} \text{aging racked up} \\ \text{on a faraway clock;} \\ \text{i.e., passage of global time} \end{pmatrix}$$

The difference between the two measures of time is totally negligible, or so at least is one's first impression on looking at the numbers: twice the mass of the Earth (2 × 0.444 centimeters) equals 0.888 centimeters. The r-coordinate of the point of observation is 6371 kilometers from the Earth's center, or 6.37×10^8 centimeters. The ratio is 1.394×10^{-9}. This fantastically small number is the fraction by which the "one" in the formula is to be diminished before we take the square root. After we have taken the square root the diminution is only half as great—that is, 6.97×10^{-10}. Not much! But in the stretch of time between the arrival of the start-of-run signal and the end-of-run signal, this factor accounts for the loss of 697 ticks out of one trillion. That 697 loss of ticks, if it could be measured so easily, would be indisputable evidence for this influence of place on one of the measures of time.

A loss of 697 ticks means a decrease in aging of 697 ticks. A difference between two clocks as small as 697 ticks out of 10^{12} lies well within today's power to measure. So far, however, science lacks any *direct* check of this full 697 figure. The reason is simple. No way has been found to station a high-precision atomic clock very far away, truly at rest relative to the center of the Earth. It has proved easier to get high precision by measuring a smaller difference. In the early 1960s, Pound, Rebka, and Snider measured the difference in rates of identical atomic clocks located at the top and bottom of a 22.5-meter shaft. They found that the higher clock goes faster than the lower clock. They measured a slowing of the rate of the lower clock, relative to the upper clock, and found it agreed within 99.9 percent with the predicted difference of 2.45 parts in 10^{15}.

What's true for the clock is true for the stone. It ages more at a higher location than it does sitting on the ground. Since maximal aging is

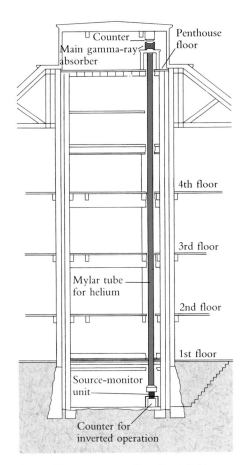

Using this arrangement, Pound and Rebka (1959) and Pound and Snider (1965) directly measured for the first time the red shift—or loss of energy—of a photon rising against Earth's gravity. Radioactive cobalt 57, here colored red, releases 14.4 thousand electron volts in photons. The photons rise 22.5 meters through a helium-filled tube (green) in a shaft at the Jefferson Physical Laboratory of Harvard University to the gamma-ray absorber (brown) and the photon counter at the top.

Labels on figure: Counter; Penthouse floor; Main gamma-ray absorber; 4th floor; 3rd floor; Mylar tube for helium; 2nd floor; 1st floor; Source-monitor unit; Counter for inverted operation its goal, why doesn't the stone rise to a great height, sit there during those 2 seconds while we're waiting on the ground, and then come back? That's a great policy for those 2 seconds, but a bad policy at the start of that 2-second time allotment and at its end. It's bad because it presupposes high speed out and high speed back. That, the Canopus adventure told us, is the way to stay young, not the way to travel the straightest possible world line, not the way to rack up maximal aging!

Aging incurred between the stone's takeoff and its return evidently registers the competition between two influences: first, the age lengthening effect of being up there and, second, the age shortening effect of traveling there and back. We can believe that there is some best height that gives maximal aging, and so a closer look demonstrates. We compare alternate histories in the table on the following page. The stone does best, ages most, racks up the maximal increment of its wristwatch time during the 2-second lapse of our ground-based time when it rises up 4.9 meters at the halfway point of its travel. Yes, we conclude, maximal aging is indeed the way to distinguish a free-float worldline, a geodesic or "straight-line" track through curved spacetime, from any other short-run worldline that connects a specified starting event with a specified terminal event. And the same principle of maximal aging governs the motion of the planets, Earth, and the Moon.

Planetary Motion

The stone is told to go into free-float here and now and yet to be back here 2 seconds from now. Given these orders, the stone picks out, from among all the up-down options open to it, the one we know. It rises up 4.9 meters and comes back down. A planet orbiting around the Sun confronts twice as many options in its travel from event A in spacetime to event B. The planet has to know not only at what instant it will best arrive at each r-coordinate circle on its way from A to B, but also how far around the Sun it will have best progressed along that circle. This double requirement on the choice of orbit has a double consequence.

First, the moving object sweeps out in each second of its wristwatch time the same effective area as in every other second of its motion (see illustration on page 170). It does not matter that sometimes the orbiting object may, like a comet, swoop in at high speed past the Sun and yet at other times sweep out at far greater distance from the Sun but at far lower speed. The effective rate of sweeping out area remains identical at the two stages of motion.

The finding has received a standard name, "the law of conservation of angular momentum." I like to illustrate angular momentum by closing a door. I can close an open door in a second by giving it a modest kick at its outer edge. But when I apply my kick only a third of the way out from the hinge to the edge, I have to kick three times as hard to close the same door in the same time. The word for what the kick gives to the door is *angular momentum*. To figure angular momentum, it is enough to multiply the linear momentum—the push of the kick—by the distance from the hinge on the door to the line of travel of my foot. It is appropriate to think of that distance as a lever arm. The linear momentum of the planet multiplied by its distance from the Sun measures the angular momentum of the planet about the Sun.

What my foot gives up when it hits the door, the door must acquire. That's conservation of angular momentum, and it applies to both the closing door and an orbiting planet. The door ends up with the same angular momentum whether my foot imparts to it a little linear momentum with a big lever arm or a big linear momentum with a little lever arm. Similarly, the planet retains an ever-fixed angular momentum throughout its orbit. Otherwise stated, the planet speeds up and slows down as it approaches and retreats from the Sun, in such a way that it sweeps out equal effective areas in equal times.

Alternate histories for a stone tossed straight up

Highest distance reached above Earth (meters)	0	2.45	4.90	7.35	9.80
Time between throw and catch as registered by a faraway clock (meters of time)	6.00×10^8	6.00×10^8	6.00×10^8	6.00×10^8	6.00×10^8
Decrease in aging at surface of Earth compared to a faraway clock (meters of time)	-0.418	-0.418	-0.418	-0.418	-0.418
Increase in aging by average higher elevation above Earth during throw (meters of time)	0	10.7×10^{-8}	21.4×10^{-8}	32.0×10^{-8}	42.8×10^{-8}
Decrease in aging incurred by up-down movement (meters of time)	0	-2.7×10^{-8}	-10.7×10^{-8}	-24.0×10^{-8}	-42.8×10^{-8}
Net increase in aging because of both higher elevation and up-down movement during the available 2 seconds (meters of time)	0	8.0×10^{-8}	10.7×10^{-8}	8.0×10^{-8}	0

A comet sweeps out equal effective areas in equal times.

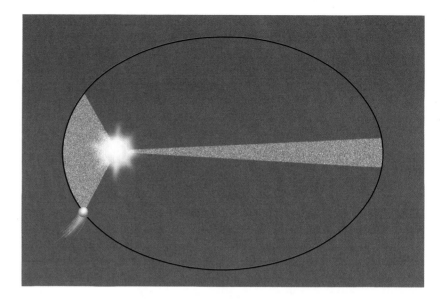

Of all consequences of this constancy of angular momentum, the most important is the speed-up of the transverse velocity that takes place when the planet comes close to the center of attraction. We can think of the motion of a planet as having two components: the planet moves directly toward or away from the Sun in in-and-out motion at the same time as it moves in a direction perpendicular to that in-and-out motion. The two straight motions combine to create a curved orbit. It is the perpendicular, or transverse, motion that speeds up near the Sun. The energy required for this speed-up has to come from somewhere. But where? This question brings us to the second consequence of the principle of maximum aging: the conservation of energy—and especially the role that angular momentum plays in the count of energy.

The principle of conservation of energy gives the quickest of all ways to describe—and understand—Newton's findings on the motion of a planet or a comet. His followers did not take long to point this out. To them the bookkeeping of energy was almost obvious. As they conceived it, energy comes in two forms. The first is kinetic energy, which depends on speed but not on position, the second, potential energy, which depends on position but not on speed. When a moving object has climbed out against what they thought of as the so-called gravitational pull of the Sun, then, said these expositors of Newton, it has accumulated a large stockpile of potential energy. Since the total energy is fixed, this accumulation of potential energy comes at the expense of kinetic energy. Kinetic energy thus depletes. As Newton's followers conceived it, however, the stockpile of kinetic energy consists not of one pile of

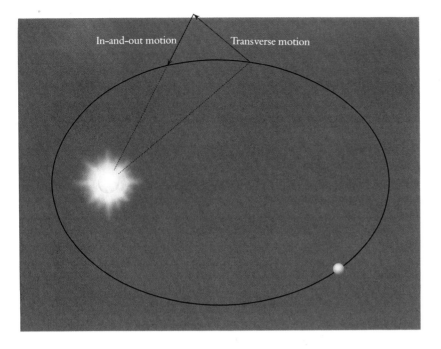

In-and-out motion   Transverse motion

The motion of a planet has two components; motion directly toward or away from the Sun (left arrow) and motion perpendicular to that in-and-out motion (right arrow).

coins but two: kinetic energy of in-and-out motion and kinetic energy of motion transverse to the in-and-out direction. But the moving object is not free to draw on this second stockpile of kinetic energy. The law of conservation of angular momentum imposes an unbreakable contractual lien on it. So the speed of in-and-out motion has to pay the price. It does so by falling, ultimately to zero, as the object reaches the innermost and outermost extremities of the orbit. By total sacrifice of in-and-out kinetic energy, the planet manages to pay the price of its high potential energy at the outermost extremity, and of its high kinetic energy of transverse motion at the innermost extremity. Occasionally an orbit like Earth's presents itself, which is so nearly circular that the part of the kinetic energy associated with in-and-out motion is exceedingly small compared with both the potential energy and the kinetic energy of transverse motion. The Earth reaches a distance of 152.10 million kilometers from the Sun at the farthest point of its orbit and 147.10 million kilometers at its closest, giving only 5 million kilometers of in-and-out motion between the two extremes.

To account for this variation of in-and-out speed with position was one of the marvels of the Newtonian theory of gravity. But it was the still greater marvel of Einstein's geometric theory of gravity to make this account come out really right. And it has a beautiful simplicity: it lets us understand all that looping through space and all that in-and-out motion as the straightest possible line through curved spacetime.

Stones in Flight and Planets in Orbit

This modern theory predicts a fantastic new feature of satellite motion, a feature that stands totally at odds with the Newtonian prediction that an object endowed with any angular momentum at all will never crash into a point center of gravitational attraction. According to the old view, the satellite will approach the center of attraction at high speed, loop around it at still higher speed, and then either go off forever or, if it has less energy, go off until it reaches a maximum separation and return to repeat again and again its death-defying swoop. The geometric view of gravity, in contrast, holds out a richer range of options. Endowed with a low enough angular momentum or a high enough energy or both, the moving object will always be sucked in and "die the death." Moreover, between motion of this catastrophic character and motion closer in character to that of the older Newtonian prediction, there is a remarkable divide. Endowed with one or another critical combination of energy and angular momentum, the incoming object will approach slower and slower to a certain well-defined r-coordinate, never quite attaining it even in infinite time, yet in the attempt forever whirling round and round in orbit. Endowed with the same angular momentum but a trifle less than the critical amount of energy, the moving object in a finite time will conclude, in effect, that it can't make it over the barrier that stands in its way. After ten turns or a hundred or a thousand the object will retreat to the great distances from which it came. On the other hand, endowed with an energy a little greater than the critical amount, after many orbits the moving object will make it over the barrier and plunge into the center of attraction in a quick swoop.

A more quantitative account of orbiting springs straight from Einstein's formula for conserved energy, which in turn derives from the principle of maximum aging. The significance of this formula shows itself nowhere more clearly than in a diagram for effective potential, or more descriptively "energy hill." The diagram shows the white plaster of Paris model of the energy hill that sits in the science museum of today's dream but tomorrow's achievement. At the very left lies the center of attraction, and distance to the right measures the r-coordinate. The height of the hill, at whatever distance from the center of attraction we choose to examine it, governs the portion of the total energy consumed— at that point in the motion—by two nonnegotiable demands for energy. In Newtonian language, one of these demands is the energy required to climb out to that r-coordinate against the pull of gravity; the other is kinetic energy of the transverse motion required to keep the angular momentum up to scratch. However, the modern Einstein geometric account of planetary motion shows that these two demands, accurately analyzed, combine in a way more complex than strictly additive to fix the height of the hill. (For details, see box on pages 174–175.)

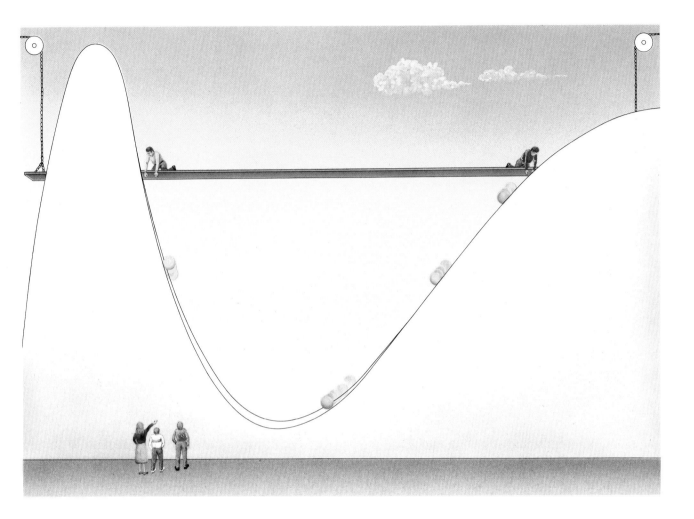

The museum guide directs our attention from the white plaster of Paris energy hill to the other factor in the motion, the total energy available. He shows us how to fix it by turning the knob at the right of the giant showcase. It raises or lowers the platform, the height of which—he tells us—represents launch energy. He invites us to push the button and launch the motion from the lowest platform level. The green billiard ball slowly rolls out from the launching platform onto the white hill. The platform, however, is so low that the ball rolls back and forth only a little either side of the minimum in the curve. The guide interprets the spectacle, "The blackness to your left ends with the blackness of a black hole.

The left-right motion of the billiard ball on the hard plaster of Paris surface symbolizes changes in distance between satellite and center of attraction, as the satellite moves closer in or farther away.

THE "ENERGY HILL"

As a planet orbits a spherically symmetric center of attraction, its around-and-around motion goes faster when close in, slower when far out, but always in the same direction. The motion in the in-and-out direction, however, is forever reversing. This change in r-coordinate goes fastest in mid-course. It slows in nearing the outer and inner extreme r-values of the orbit, as if there the planet were having to climb a hill. All these features of the motion plus a quantitative description of this "energy hill" follow from a single simple principle: a free-float object makes its way up through curved spacetime along the straightest worldline it possibly can. Spacetime, however, stands in striking contrast to space. Any local departure from the ideal geodesic would bring about, not an increase, but a decrease in "aging," of which the kinked roundtrip from Earth to Canopus and back, in Chapter 3, provided an extreme example (40 years vs 202 years).

A local departure from maximal straightness would result from a mistaken choice of the angular coordinate, or of the time coordinate, at which the object arrives at a specified r-coordinate midway between two nearby specified events A and B on the worldline. Therefore, the first requirement for maximum straightness of route through the curved Schwarzschild spacetime is the right choice of *angle* for the intermediate event, the choice required to maximize the aging incurred between A and B. It turns out that the age-maximizing choice makes a quantity known as the "angular momentum per unit mass" the same on the two stretches of the worldline from event A to intermediate event and from intermediate event to event B. Therefore by extension of the argument, the correct choice of angle attained at every r-value makes the angular momentum constant through the whole course of the worldline:

$$\text{angular momentum per unit mass} = \frac{r^2 d(angle)}{d(proper\ time)} = \text{constant}$$

Here r is the Schwarzschild r-coordinate of the moving object (idealized as a point), and $d(angle)$ is the very small angle the object moves through around the center of attraction during any very small lapse, *d(proper time)*, of its own planetary wristwatch time. The second requirement, the age-maximizing choice of *time* for arrival at this intermediate *r*-value, makes the "energy per unit mass" the same on those two stretches on the worldline and therefore everywhere:

$$(\text{energy per unit mass})^2 = [dr/d(proper\ time)]^2 + r\text{-dependent height of hill}$$
$$= \text{constant}$$

Here "height of hill" is an abbreviation for a quantity that depends on the mass *m* of the center of attraction and on the planet's *r*-coordinate and on its angular momentum per unit mass, but not on time and not on speed:

$$\text{height of hill} = (1 - 2m/r)\left[1 + \frac{\left(\dfrac{\text{angular momentum}}{\text{per unit mass}}\right)^2}{r^2} \right]$$

The diagram on page 176 depicts this height of hill and its dependence on *r* in the case when the angular momentum per unit mass—a quantity measured in meters when we use geometric units—happens to be 3.75 times as great as the mass of the center of attraction. It also shows, by fainter curves, the energy hill for other sample values of angular momentum. The platform height is our pictorial word for the square of the energy per unit mass. That energy per unit mass is, moreover, the ratio of two quantitive measurements in meters, and thus is itself a pure number independent in its value of choice of units. The two quantities, energy and angular momentum, both expressed on a per-unit-mass basis, suffice completely to distinguish the shape of one planetary orbit from another, as the diagram illustrates: the green orbit for the choice of platform height or (energy)^2 indicated by the green line; and likewise for the yellow, orange, and red platform heights and orbits. The red orbit and central black hole are depicted on an enlarged scale. All orbits were calculated especially for this book by Christopher A. Mejia, a student of Dr. Edwin Taylor.

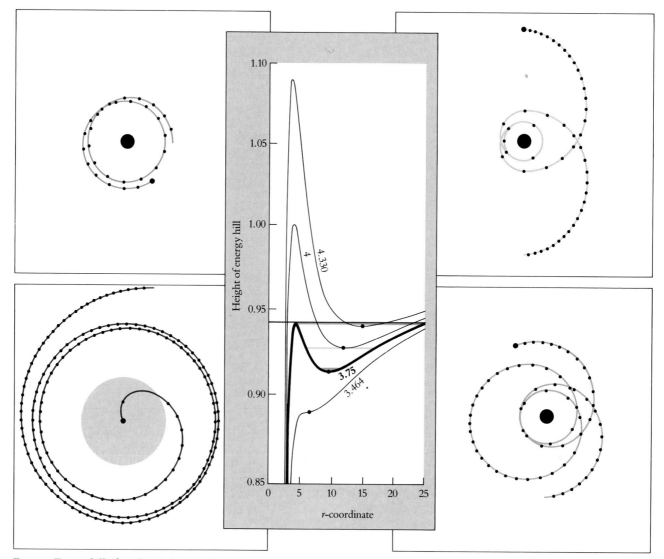

Center: *Energy hills for selected choices of angular momentum.* Outlying diagrams: *The paths of satellites orbiting in the field of the center of attraction. All the satellites have an angular momentum per unit mass of 3.75. Their orbits therefore differ according to their energy per unit mass, which is color coded to the height of the energy platform in the central diagram. The dots on each orbit indicate the passage of equal intervals of proper time. The center of attraction is a black hole, whose size is indicated by the central black circle. The black hole's radius is 2m, where m is its mass. The distance scale for the* r-*coordinate in the central diagram is expressed in units of the mass of the center of attraction (14.7 kilometers for a black hole of ten solar masses). Not shown are circular orbits, with no in-and-out motion at all, corresponding to the black dot at the bottom of the energy hill. The diagram at lower left is shown at an enlarged scale so that we can follow the orbit as it plunges to the central point of crunch. What a way to correlate in-and-out motion of the satellite—pictured as the rolling of a ball left and right on the energy hill—with round-and-round motion—governed by Kepler's law of equal area in equal time!*

The billiard ball's left-and-right oscillation tells us something of the motion of a comet that has fallen under the influence of that black hole. It simulates on a reduced scale that comet's in-and-out motion in its not quite circular motion about the black hole. To be sure, the billiard ball reveals nothing of the comet's round and round motion, but that lack is not important because we know that angular momentum stays constant, and out of angular momentum and distance from the center we can always predict the angular velocity or the velocity of transverse motion. So everything we need to predict an orbit, we get entirely out of our knowledge of the energy hill."

We turn the knob, raise the platform, and release a new billiard ball from a higher level. This yellow ball (shown in the diagram on page 173) zooms downhill to the left, climbs the lefthand hill, slows, and turns back about a third of its starting distance from the left. The orbit to which it corresponds swoops in at its closest to only a fraction of the r-coordinate of the orbit's outer extremity. If the Newtonian vision of gravity were the correct one, time to go once around in orbit and time to go in and out in orbit would so miraculously coordinate that the orbit would be an ellipse. In Newtonian theory, however, the energy hill has a shape different from what we see before us in the diagrams. There is no pit in the hill close to the center of attraction and no summit to the hill on the left. It climbs and climbs and climbs. In Newtonian theory, the shape of the energy hill is a miracle. It guarantees that the time for one in-and-out oscillation always agrees exactly with the time for one full 360-degree or 2π turn of the orbit. In the geometric theory of gravity, however, the energy hill displays a barrier summit, and inward of it, a pit. The Einstein hill, by thus not climbing into the sky, declines on the left from the Newtonian one. The billiard ball finds the summit more inviting than it would Newton's upward climbing extension of the curve. It spends more time there—before turnaround and return—than Newtonian theory predicts. It takes more than the time for one 360-degree turn to go from outer extremity of the orbit to inner extremity and return. The curve does not close. Otherwise stated, the would-be ellipse of Newtonian theory slowly rotates; its axis "precesses."

The mass of the Sun, expressed in geometric units, is 1.47 kilometers, and the foot of the pit in the energy hill lies at an r-coordinate of twice this amount, 2.94 kilometers. By comparison, the planet closest in, Mercury, varies during its 88-day year from a minimum r-coordinate of 46.0 million kilometers to a maximum of 69.8 million kilometers. Even Mercury never comes anywhere near so close to the barrier summit or the pit in the energy hill as the museum's plaster of Paris model would suggest. The reason is simple—the designer picked extreme conditions. Nevertheless, Mercury, far from the barrier summit as it is, still "sniffs

FROM KEPLER'S LAW TO NEWTON'S GRAVITY

Shortly before his death, the great Danish astronomer Tycho Brahe (1546–1601), then working in Prague, assigned to his new assistant from southern Germany, Johann Kepler (1571–1630), the task of making sense of the motion of Mars, more resistant to simple description than any other planet. No one could have been a better choice. Kepler had the perserverance to do enormous calculations day after day in the passionate belief that nature is ruled by the simplicity and beauty of geometry. Guided by this belief, he discovered three regularities of planetary motion that have become eternally associated with his name:

■ *Kepler's first law:* A planet moves about the Sun in an elliptic orbit, with one focus of the ellipse located at the Sun.

■ *Kepler's second law:* A straight line from the Sun to the planet sweeps out equal areas in equal times.

■ *Kepler's third law:* The time required for a planet to make one orbital circuit, when squared, is proportional to the cube of the major axis of the orbit.

These findings—destined to be so decisive for astronomy, for the understanding of gravity, and indeed for the whole science of motion—were buried in such a mass of other findings, partly developed theory, and speculation that their impact took long to make itself felt. Isaac Newton, omnivorous reader that he was, would probably never have seen Kepler's laws had he not come upon a summary of them in the book *Astronomia Carolina* by Thomas Streete. They became a cornerstone of Newton's theory of the heavens.

Hidden in Kepler's first law is another law: for a planet traveling in a Sun-focused ellipse, the central force acting on a planet must fall off inversely as the square of its distance from the Sun. Amazingly, Kepler himself already in his own writings spoke of the possibility of an inverse-square force acting on a planet. However, he had the mistaken motorboat notion of force: no motor, no motion. Not yet to him nor to anyone was it clear that a mass will continue to move in what is to it a straight line through space. How could he realize that no force is needed to drive the planet *around* its orbit, only a Sun-centered force to hold it *in* orbit (Newton) or a Sun-centered spacetime curvature (Einstein)? To make that step forward the world needed the concept of inertia, fruit of the investigations of Galileo and Newton.

Johann Kepler

Born December 27, 1571, Weil, Württemberg, Germany. Died November 15, 1630, Ratisbon, Germany.

I measured the skies
Now the shadows I measure
Skybound was the mind
Earthbound the body rests.

Johann Kepler wrote his own epitaph, here translated from the original Latin by J. A. Coleman.

Kepler himself gave us the most powerful proof that the force acts toward the Sun, not in the direction of the planet's motion. He gave us this message in his second law, undeciphered though it was by Kepler himself or any contemporary. It is not surprising that this secret lay hidden, as not one word about force occurs in his second law: each planet, traveling in its elliptic orbit around the Sun, sweeps out equal areas in equal times.

To us, the message unfolds without difficulty. Inertia is the secret key. As we have already seen, to say that a planet in its motion around the Sun sweeps out equal areas in equal times can be translated into a powerful and familiar message: the angular momentum of the planet around the Sun does not change from moment to moment. Inertia guarantees conservation of angular momentum in the absence of any round-and-round force pushing on the planet. Not one bit of round-and-round force acts on the planet. It is no motorboat that needs an engine and propeller to drive it through space. The only source of guidance it has is Sun-centered.

To translate Kepler's third law of motion into a statement about the force acting on an orbiting planet is more difficult than in the case of the first and second laws. Nevertheless, Newton succeeded in doing so. His "translation," expressed in geometric units of mass, length, and time, is simple indeed:

$$\begin{pmatrix} \text{mass of} \\ \text{center of} \\ \text{attraction} \end{pmatrix}^1 = \left(\frac{2\pi}{\text{time for one orbit}} \right)^2 \times \begin{pmatrix} \text{half of} \\ \text{the long axis of} \\ \text{the elliptic orbit} \end{pmatrix}^3$$

Of all the ways to determine the mass of a heavenly body, the most reliable, easiest to apply, and most used derives from this formula. In accord with Charles Misner and Kip Thorne, I call the formula the "Kepler 1-2-3 law." Here it is restated in terms that are relativistically correct:

$$\begin{pmatrix} \text{mass of center} \\ \text{of attraction} \end{pmatrix}^1 = \begin{pmatrix} \text{faraway time satellite} \\ \text{requires to cover one} \\ \text{radian of its circle} \end{pmatrix}^{-2} \times \begin{pmatrix} r\text{-coordinate} \\ \text{of the circle} \end{pmatrix}^3$$

The Kepler 1-2-3 law, expressed in this form, applies to a circular or almost circular orbit around a far more massive center of attraction. In

(Continued on next page.)

Calculating mass by applying the 1-2-3 law

Satellite	Time to cover 1 radian of orbit	r-coordinate (meters)	Mass of center of attraction (meters)
	Calculating the mass of the Sun		
Earth	58.13 days or 1.506×10^{15} meters	1.496×10^{11}	1.477×10^{3}
Jupiter	689.55 days or 17.86×10^{15} meters	7.783×10^{11}	1.477×10^{3}
	Calculating the mass of Jupiter		
Io	0.282 days or 0.729×10^{13} meters	0.422×10^{9}	1.41
Europa	0.565 days or 1.464×10^{13} meters	0.671×10^{9}	1.41
Ganymede	1.139 days or 2.95×10^{13} meters	1.070×10^{9}	1.41
Callisto	2.656 days or 6.88×10^{13} meters	1.883×10^{9}	1.41

this formula, mass, time, and distance are all measured in the same units, here meters. Likewise, the r-coordinate can be identified with the speed of approach to us when the satellite is seen edge on multiplied by the time it requires to cover 1 radian of its orbit. The table shows some sample applications of the 1-2-3 law. First is a check that this law, applied to two different planets, Earth and Jupiter, gives the same mass for the Sun. Likewise applied to the satellites of Jupiter, the 1-2-3 law yields in all four cases the same mass for the center of attraction, Jupiter itself. It is a marvelous consequence of this 1-2-3 law that the time to go around a sphere in a surface-hugging orbit is independent of the size of that sphere and dependent only on its average density. A satellite takes 84 minutes to orbit once around the Earth, 84 minutes to orbit once around an asteroid of the same average density of the Earth, and 84 minutes to orbit once around a spherical rock of that same density. In brief, density governs spacetime curvature, and spacetime curvature governs time in orbit.

Io (blue), Europa (green), Ganymede (yellow), and Callisto (pink) followed for 14 days in December, 1964, relative to Jupiter (red bar).

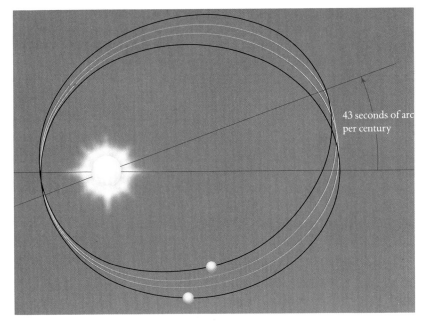

out" that the region of the energy hill on which it oscillates in and out is not exactly Newtonian. In consequence, Mercury's orbit does suffer a small non-Newtonian precession. Its predicted amount is only 43 seconds of arc per century.

Evidence for the precession of the orbit of Mercury had already accumulated before 1915 and Einstein's general theory of relativity. It came from the great American astronomer Simon Newcomb (1835–1909). As Kepler, with his passion for beauty, order, and mathematical precision had smoked out the ellipse hidden for so many centuries, so Newcomb—the conscience of planetary motion in his own day—brought new tools of observation and calculation into service. For the other astronomers of his day, as for Newcomb himself, it was not enough to consider each planet in turn in isolation from the others. The enormous labor, however, of trying out this and that mass value for each planet and then figuring out the resultant motions of it and all the planets—the only really thorough-going check of theory against observation, and observation against theory—was too overwhelming for anyone to undertake except a man of Newcomb's ability and strength of character. By 1898 he had concluded that in the course of a century of 88-day orbits, the major axis of Mercury's elliptic orbit turns through 43 seconds of arc more than can be explained by any known effect. Though the angle was small—the angle subtended by a child's foot a kilometer away—Newcomb's investigations made it into a rock of astronomy, well-established and incontrovertible. This discrepancy was widely known by 1915, and various implausible explanations had been put forward to account for it. Each proved wanting and was discarded. No believable explanation for this precession ever turned up until, miraculously and quite unbidden, Einstein's geometric theory of gravity predicted it. Of all early tests of the idea of curved spacetime this was the earliest, the most convincing, and the most dramatic.

By now we have grown impatient and twist upward by another notch the dial that controls the billiard ball's level of launch. The old ball is automatically rolled off and out of sight and a new, orange one starts from a new and higher level. It zooms down through the minimum in the energy hill, which symbolizes that place where the effective kinetic energy of the inward directed motion of the comet is greatest. Then that energy starts to decrease; otherwise put, the orange ball climbs up the energy hill and slows down as it does. The ball almost makes it over the top but not quite. While its teetering in indecision whether it *is* going to make it, the corresponding comet is orbiting round and round. Ultimately the ball turns back and retraces its route. On each point of the outbound trip it reaches an outbound speed equal in magnitude but opposite in direction to the speed it had when inbound. The orbit has the shape depicted by the orange line in the diagram on page 176.

Realizing how close to critical the adjustment was that we made with that last turn of the dial, this time we turn only a smidgen further, to the position so carefully marked as "critical." The orange ball plops out of sight, and from the new and only slightly higher position of the platform a new ball starts its course downward from the right. It zooms through the minimum of the plaster of Paris model of the energy hill and climbs up the left-hand slope, moving ever slower as it does. A whole minute goes by and still it's creeping, hardly visibly but still creeping. We're prepared to believe that it will sit there teetering at the summit indefinitely if not disturbed when, oops, underground a subway train roars through, shakes the building, and just then the lights go out and it already being five minutes past closing time the outcome remains in doubt. Will the ball plunge down the wall of the pit onto the center of attraction or gather speed outward and ultimately repeat the cycle? Next morning we are back to see the denouement. This time we adjust the platform level a mite past critical. The launcher lets free a red ball. Down the right-hand hill it goes and up the left-hand one. It slows, slows greatly as it approaches the summit—allowing the corresponding orbit to make many turns—but still it goes over and plunges down the wall of the pit to destruction, symbolized in that museum's clever exhibit by a flash of light, a shower of sparks—and disappearance of the ball.

Our brief travel through the land of planetary motion reveals the power of three principles. First, the principle of extremal aging, which from motions between spacetime event A and spacetime event B selects the actual from the conceivable. Second, the law of conservation of angular momentum, which, a deeper analysis than ours would reveal, is a consequence of the spherical symmetry of spacetime geometry in which the motion proceeds. Third, the law of conservation of energy, which, as a like deeper analysis would also reveal, depends for its existence on the fact that this spacetime geometry remains the same, from one moment to the next, over the entire epoch of the comet's flight through it. Each of these principles is interesting in its own right. Still more interesting is the shape of the energy hill—and the summit of that hill—that they conspire to produce. Most interesting of all is the possibility to climb over this energy hill and plunge down the pit beyond it—a plunge to destruction which an unhappy choice of angular momentum and energy makes inescapable.

Of the "shower of sparks" predicted to be thrown out in this act of destruction, none looks more fascinating to seek, and someday find, than gravity waves.

11

Gravity Waves

Oh, Gravity Wave,
No searcher has yet beheld you,
The Golden Fleece of our day.
Courageous adventurers seek you
With all the shrewdness of Jason,
With all the witchcraft of lasers and electronics.

Why are you so much more elusive
Than your younger brother,
The electromagnetic wave?
It, like you, is the winged carrier
Of invisible messages;
Its tidings we can catch,
As news, pictures, urgent calls,
And happenings on the surface of a distant star.

Your report of cataclysms
Far down in the interior of stars,
Of turmoil in the deep and secret places of the universe,
And of the whirl of great masses
Caught in each other's tight embrace,
This very moment you must be carrying past us
Who cannot yet hear your words.
Give us soon, oh, Gravity Wave,
Our first whisper of your golden messages.

A pulse of gravity radiation (symbolically depicted) is generated by the collapse of a slowly rotating star core to a rapidly rotating neuron-star pancake.

In the depths of an ill-fated, collapsing star, billions upon billions of tons of mass cave in and crash together. The crashing mass generates a wave in the geometry of space—a wave that rolls across a hundred-thousand lightyears of space to "jiggle" the distance between two mirrors in our Earthbound gravity-wave laboratory.

A cork floating all alone on the Pacific Ocean may not reveal the passage of a wave. But when a second cork is floating near it, then the passing of the wave is revealed by the fluctuating separation between the two corks. So too for the separation of the two mirrors. There is, however, this great difference. The cork-to-cork distance reveals a momentary change in the two-dimensional geometry of the surface of the ocean. The mirror-to-mirror distance reveals a momentary change in the three-dimensional geometry of space itself.

The idea is old to extract energy from ocean waves. After all, the ability of a water wave to change a distance lets itself be translated into the ability to do work. The same reasoning applies to a gravity wave. Because it can change a distance, it can do work. It carries energy. Mass-energy once resident in the interior of a star has radiated out to us and to all the universe.

Of all the workings of the grip of gravity, none is more fascinating than such a gravity wave, or opens up for exploration a wider realm of ideas. None pushes to a higher pitch the art of detecting a small effect, and none gives more promise of providing an unsurpassable window on cataclysmic events deep inside troubled stars. Nevertheless, no other great prediction of Einstein's geometric theory of gravity stands today so far from triumphant exploitation. As of this writing, not one of the nine ingenious detectors built to this day has proved sensitive enough to secure any generally agreed detection of an arriving gravity wave.

Why Does Mass Lose Energy?

Ever since the days of Niels Bohr, the hallmark of the Copenhagen Institute for Theoretical Physics has been an interest in deep puzzles of principle. Therefore it is not surprising that two members of that institute, Jørgen Kalckar and Ole Ulfbeck, should have set themselves this question: Does any truly simple line of reasoning assure us that gravity will— or will not—inescapably carry energy away from two masses that undergo rapid change in relative position? I'd like to translate their thinking and their "yes" conclusion into a story that savors of mythology. The Atlas of our day, zooming through space in free float, insists as much as ever on maintaining physical fitness. He pumps iron, not by raising iron against the pull of Earth's gravity, but by throwing apart two identical great iron spheres, Alpha and Beta. He floats between those minor moons and plays catch with them. Each time they fall together under the influence of their mutual gravity, he catches them, absorbs their energy of infall in his springlike muscles, and flings them apart so that they always travel the same distance before returning. It's an enchanting game, but Atlas finds that it's a losing game. When the masses fly back together, they never yield up to him as much energy as he must supply to throw them apart again. Why not? Kalckar and Ulfbeck explain.

Their central point is simple: time delay. Like any force that makes itself felt through the emptiness of space, the force of gravity cannot propagate faster than the speed of light. This limitation imposes a delay on the attraction between the two iron spheres. Alpha, on each little stretch of its outbound path, feels a pull that originated from Beta when the two were a tiny bit *closer* than they are now. The actual force that's slowing Alpha is therefore a tiny bit bigger than we would judge from thinking of them as stationary at their momentary separation. On its return trip inbound along the same little stretch of path, Alpha experiences a helping pull that originated from Beta when the two had a separation slightly *greater* than its present value. The actual force that's speeding Alpha inward is therefore a tiny bit less than we would judge from thinking of them as stationary at their momentary separation. In each stretch of their outbound trip, the two masses have to do more work against the pull of gravity than they get back—in the form of work done on them by gravity—on the same stretch of path inbound. A calculable amount of energy disappears from the local scene on each out-in cycle of Atlas's exercise. Yet total energy must somehow be conserved. Therefore the very gravity that steals energy from Atlas and his iron, or from any two masses that rapidly change their relative position, must somehow all the time be transporting that stolen energy to the far away. That inescapable theft of energy, Kalckar and Ulfbeck conclude, is in its quality, its directional distribution, and its magnitude none other than what Einstein had treated long before under the head of *gravity radiation* and what we now call gravity waves.

A Gravity Wave Is Transverse

The whole concept of gravity waves only slowly won acceptance. Indeed, in the 1940s and 1950s, several able scientists even denied that there could be any such thing as transport of energy by gravitational radiation. Investigators—among them some who initially had been skeptical—gradually came to terms with two central features of a gravity wave. First, a gravity wave is *transverse,* not longitudinal. A gravity wave truly reveals itself in directions perpendicular to—that is, transverse to—the direction of its travel. In those directions, space alternately squeezes and stretches. For example, for a gravity wave traveling "upward," the stretch of space at a given point and at a given instant may be greatest in the left-right direction, and the shrinkage at the same instant and point greatest in the in-out direction. In contrast, space in the direction of travel of the radiation alternately squeezes and stretches by a totally negligible amount. In brief, a gravity wave is a traveling disturbance in the geometry of spacetime that acts at right angles to, or transverse to, the direction of propagation. Second, however far this traveling region of spacetime curvature has progressed, there the measure of the deformation it makes in spacetime geometry has fallen off in proportion, not to the inverse square of the distance of travel away from the source, but to the *inverse first power* of this distance. Of two observers located in the same direction "above" the source, one twice as far away as the other, the one more remote will experience half the left-right stretch of the other and half the in-out squeeze.

Both features of a gravity wave—transversality and slow falloff of strength with distance—created in the 1940s and 1950s a bewilderment that echoed the discomfort investigators had experienced half a century earlier with the similar features of an electromagnetic wave. Chief among doubters had been the great Scottish investigator William Thomson, also known as Lord Kelvin. This pioneer of the trans-Atlantic cable, originator of many of our units and principles of measurement in the field of electricity and magnetism, and author of the theory of waves guided by wires, had trouble accepting the central idea of electromagnetic waves. That there are electromagnetic waves, Kelvin could hardly doubt. Already in 1887 Heinrich Hertz had discovered the propagation of electromagnetic action through space at laboratory distances. And on December 12, 1901, Guglielmo Marconi, on his first attempt, succeeded in transmitting and receiving signals across the Atlantic Ocean. But Lord

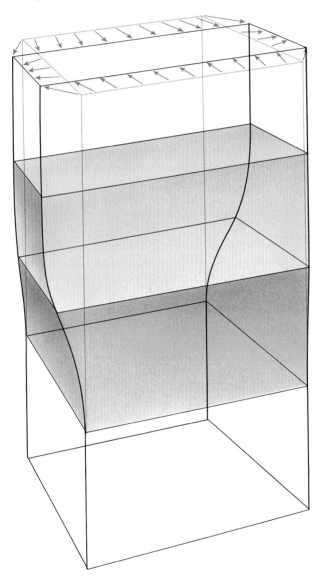

Tiny test masses respond to a pulse of gravitational radiation by changing the path of their worldlines, the thick red lines heading upward. The particles don't change their separation in the direction of travel of the wave, but in one transverse direction are first pulled apart, then pushed together, and in the other transverse direction are first pushed together, then pulled apart.

Kelvin long questioned that existing theory accounted for those waves. In his book in 1904 based on his famous Baltimore lectures, he wrote, "The so-called 'electromagnetic theory of light' has not helped us hitherto . . . it seems to me that it is rather a backward step . . . the one thing about it that seems intelligible to me, I do not think is admissible . . . that there should be an electric displacement perpendicular to the line of propagation."

Really to understand a new idea, we must fight it. Otherwise we are in the position of those of whom G. K. Chesterton remarked, "It isn't that they don't agree with the answer. They don't even understand the question." Or, if I may abbreviate and oversimplify Henry James, "To hate is to distrust. To distrust is to probe. To probe is to study. To study is to appreciate. To appreciate is to defend. To defend is to love." For Lord Kelvin to be upset with the central idea of an electromagnetic wave was the first step in his coming to both know and exploit such waves. To sketch the struggle to understand an electromagnetic wave is to travel the simplest route that I know—and the road I traveled myself—to an understanding of the how and why of a gravity wave.

The Kink in Electromagnetic Waves

As a 10-year-old Vermont schoolboy, I was fascinated by the delights and mysteries of electricity, as this subject was revealed by the farmhouse copy of J. Arthur Thompson's *Outline of Science*. Of all the mysteries in the subject, none was more puzzling than the reach of radiation. For a country boy like me, it was natural to picture electric lines of force as sticking out from a center, like quills sticking out from a porcupine's body or, more terrifyingly, thrown out by a twitch of the porcupine's tail. A square of tissue paper one meter from the dangerous porcupine may be pierced by 20 quills, but held out at twice the distance from the same discharge, it will be pierced by only 5 quills. The reason was evident. The square of tissue paper has not changed in size. However, it is competing for quills with an imaginary spherical surface. At twice the distance the radius of this sphere has doubled, and therefore its surface has increased fourfold. The square of tissue encompasses a four-times smaller fraction of the surface of this larger sphere than of the smaller one. This picture made it easy for me to accept the proposition that the electric field of force produced by a localized electric charge falls to a quarter of its value when the distance from the charge is doubled. I verified this idea by doing experiments in the crackling dry air of the Vermont winter with the tip of an electrified comb acting on tiny bits of paper. To discover that the inverse-square law applied to the action of an

electric charge, and to learn that Newton had successfully invoked the same inverse-square law for gravity, made the workings of the world seem totally reasonable—until I learned more!

Then I encountered in J. Arthur Thompson the statement that the electric force in an electromagnetic wave points, not straight out from the center of attraction, but transverse to that direction, and with a strength proportional to the inverse first power of the distance. This behavior of the radio wave baffled this schoolboy as much then as it had disturbed Kelvin twenty years earlier, and as much as the analogous behavior of a gravity wave was destined to puzzle and upset troubled physics colleagues twenty years later. My puzzlement grew more infuriating. Now thirteen, and living in the steel city of Youngstown, Ohio, I delivered *The Youngstown Vindicator* to fifty homes after school. A special weekly section in that paper reported the exciting developments in the new field of radio, including wiring diagrams for making one's own receiver. And my paper-delivery dollars made it possible for me to buy a crystal, an earphone, and the necessary wire. The primitive receiver that I duly assembled picked up the messages from KDKA, the world's first regular radio broadcasting station, located in Pittsburgh, Pennsylvania, a hundred kilometers away. What joy! But what raging bewilderment! According to the inverse-square law, how could electric charges or elec-

J. J. Thomson

Born December 18, 1856, Cheetham Hill, England. Died August 30, 1940, Cambridge, England. Thomson discovered the electron in 1897 and in 1906 was awarded the Nobel prize.

tric currents—even the very large ones at the disposal of the Westinghouse engineers in Pittsburgh—exert any appreciable influence at Youngstown? Whatever the strength of the electric field 1 centimeter from its source in Pittsburgh, it would have decreased by a factor of 10^{14}, according to the inverse-square law, at Youngstown, 100 kilometers or 10^7 centimeters away. Thus the strength of field would have fallen to a totally undetectable level. Yet there I was, listening in!

The mystery, the bewilderment remained for several more years. As a high-school boy, I had not yet learned relentlessly to dig out the best thinking on an issue, whatever it might be. In 1934 I was finally exposed to the best thinking in the form of Sir J. J. Thomson (1856–1940). Master of Trinity College, Cambridge, and discoverer of the electron in 1897, this short, wiry, grey-haired man was indeed at that time the grand old man of physics. After seeing and hearing Thomson at an international conference, I at last had sense enough to read his 1907 book, *Electricity and Matter*. Then the central idea of electromagnetic radiation suddenly became clear in a single word—*kink*. A kink of electric line of force—translated to the idea of a kink in the direction of a spacetime deformation—will prove the magic ingredient for understanding what a pulselike gravity wave is, and why there should be any such thing.

No picture of what happens is better than J. J. Thomson's. An electric charge travels to the left at a fixed speed and carries with it its full sunburst panoply of electric lines of force. Suddenly the charge is struck a blow and thereafter travels to the right at the same fixed speed. What does the pattern of lines of force look like at some specified instant after the whang? Say 10 centimeters of light-travel time later? Ten centimeters of time means 10 centimeters of light-travel distance. We draw a circle—symbol of a sphere—that has that 10-centimeter radius and that is centered on the point of whang. Inside that circle the lines of force have heard about the new state of affairs. They radiate straight out from what is at that instant the actual position of the charge. Outside that circle they haven't yet gotten word of the new dispensation. They radiate out from the point where the charge *would* have been at the present instant if it had continued on in its old left-directed motion at its old velocity. But what happens to the lines of force as they make their way from the one pattern to the other?

Electric lines of force have a striking feature. No electric line of force ever ends in charge-free space. Otherwise stated, every little cube of space has entering it no more lines of force than leave it. With this feature in mind, let's keep strict count of every line of force that leaves the charge, or keep account at least of enough sample lines to know what is going on. Three will do—three out of a hundred that we might have examined!

Line number zero points straight to the left after the collision as well as before it. Line number 100 points straight to the right after the collision as well as before it. No kink in either line! But look at line number 50. Inside the "sphere of influence" centered on the point of whang, line number 50 points straight up from the actual present position of the charge. Outside the sphere of influence line 50 points straight up from the would-have-been present position. If any line of force looks broken, 50 is the candidate. There seems to be no way for it to get from the one assigned outward stretch of route to the other, no way to remain unbroken, no way to fulfill the legal requirement of "no loose end." Every bit of space otherwise available for maneuver seems already to be preempted by lines of force with other numbers—within the sphere, the full sunburst of lines pointing straight out from the charge; outside the sphere, the full panoply of lines pointing straight out from the charge's would-have-been position to the faraway. The crisis signals an upcoming moment of truth. How *can* that line manage not to break its course?

Suddenly we see it. The sphere is not a sphere. It is a spherical shell. The electric charge at its center did not suffer a totally instantaneous alteration of velocity. The change took time, a time perhaps quite short, but nevertheless finite. To be concrete, let's say the velocity change took a millimeter of light-travel time, that is, 3.3×10^{-12} seconds, or 3.3 picoseconds in conventional units of time. Then the thickness of the shell is one millimeter. A shell of space of that thickness provides a conduit within which line 50 threads along, from its point of entry into the shell to its point of exit from the shell. That line of force has to make one sharp bend where it enters this conduit and a second where it exits, but be-

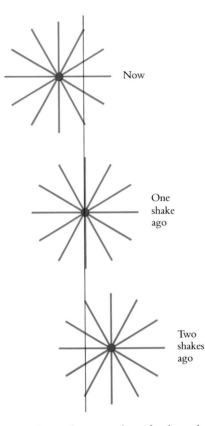

As an electric charge travels uniformly to the left, the pattern formed by its lines of force moves with it. One "shake" is the Los Alamos term for 10^{-8} seconds (from "one shake of a lamb's tail").

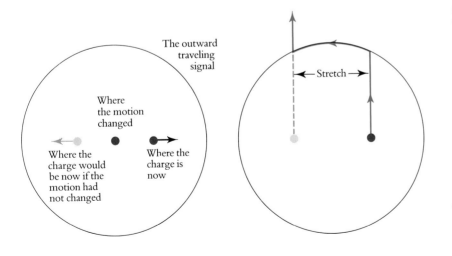

Left: *The signal that the motion of an electric charge changed is centered on the place where the motion changed.* Right: *An electric line of force "stretches" to arrive at the point where it would have been had not the electric charge changed its motion.*

A pulse of electromagnetic radiation is carried outward with the speed of light. The pulse is localized in the spherical shell or "conduit" where the electric lines of force suffer tight packing because they are kinked. These kinks, central feature of the pulse, correspond in length to the difference at the present instant between the would-have-been position of the charge and its actual position.

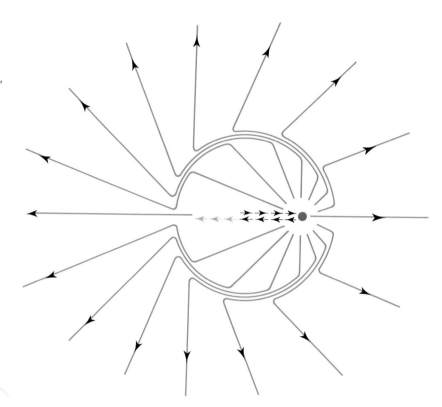

tween the two kinks it runs smooth and very nearly parallel to the course of the electric charge itself. Its length in this run is practically the same as the separation between the would-have-been position of the charge and the actual position. Its length there is many times as great as the one millimeter length it would have had if it ran straight across the shell. In its thin shell the conduit has to accommodate enormously more length of lines of force than would be the case if those lines ran straight across the shell. In the conduit is concentrated the strength of the electric field.

At last we have laid out before our eyes the workings of an electromagnetic wave of pulselike character. All the features are here. First, the lines of force inside the kink run transverse to the direction of outward propagation, the direction back to the source. Second, the strength of this transverse field is far stronger than would be implied by any direct application of the inverse-square law. Thus, as time goes on and the sphere of influence expands, so also the separation increases between the would-have-been position of the charge and its actual position. More length of lines come up, but also more length of conduit in which to accommodate them. No whit of increase takes place with advance of time in the packing density of lines as that density shows itself in the plane of the paper. Only in the other direction, the direction perpendicular to the plane of the paper, does the expansion of the sphere loosen up the packing of the lines of force. The packing density of lines of force measures the strength of the electric field. In consequence, that strength falls off only as the first power of the radius of the sphere, only as the first power of the distance from the source.

The strength of the electric field in the pulse is directly proportional to the acceleration of the charge that produces it. Thus the packing ratio, the ratio between the actual length of the line in the conduit and the straight-line distance across the conduit, is directly governed by the ratio between the velocity change of the charge and the time it takes to accomplish this velocity change—and this ratio is exactly what we mean by acceleration. No acceleration, no radiation!

Now we come to the fourth and last feature of an electromagnetic wave. The stretch of line as it passes through the conduit is greatest for line number 50, and there is no stretch at all in lines 0 and 100 as they cross the conduit. In brief, there is no fore or aft radiation, and the observer finds the radiation rising slowly to its summit of intensity in the direction totally perpendicular to the line of acceleration. No wonder that the antenna of a broadcast tower rises straight up in the air so that the slosh of electric charge up and down it can have maximum reach out across the surrounding landscape.

A Gravity Wave Compared with an Electromagnetic Wave

Almost every feature of an electromagnetic wave transcribes itself into a corresponding feature of a gravity wave. At the legalistic level, to be sure, there is a vital difference. We can describe an electric field by lines of force that never end in any region free of charge. In contrast, spacetime curvature is the descriptor for gravity, and spacetime curvatures add to zero for any three perpendicular plaquettes that, with their facing partner plaquettes, surround any small cube free of momentum and energy. Still, the Newtonian description of gravity stands nowhere in such close correspondence with electric theory as in the inverse square law of dependence of force of distance. And exactly this law was the heart of J. J. Thomson's way into an understanding of electromagnetic radiation. Out of this circumstance, we begin to see the quickest way to a first understanding of gravity waves: Adopt for a start the wrong, but useful, idea of gravitational lines of force. Then translate the conclusion into correct thinking and correct language.

After once again hurling the two great iron spheres, Atlas is suddenly called away. In consequence, those two great iron spheres do not come to their usual Atlas-cushioned stop. They crash to a halt. As we expect, that sudden alteration in the velocity of Alpha generates an outgoing transverse pulse of gravity radiation. The strength of that pulse is proportional to the mass of Alpha, proportional to the sharp momentary deceleration, inversely proportional to the first power of the distance to the point of observation, and highest at a point of observation located transverse to the line of impact. But wait! There is Beta to be considered, too. It suffers a velocity change equal in magnitude but opposite in direction to that of Alpha. It too, therefore, in our rough and ready way of speaking, generates a pulselike configuration of the gravitational lines of force, but its lines of force run directly opposite to those generated by Alpha. Conclusion: The two pulses cancel. No gravity radiation! But let's look again. Yes, we do get exact cancellation of the two oppositely directed pulses in the direction of observation exactly perpendicular to the line of approach of Alpha and Beta. And yes, along the line of approach itself, there is not even any pulse to be cancelled. But when we find ourselves at any angle intermediate between these extremes, for example at 45 degrees, we do receive pulses from both masses and they are equal in magnitude and oppositely directed. But they don't cancel! Or they don't totally cancel. The pulse from the more remote mass arrives a tiny bit later than the pulse from the nearer mass. In consequence, the observer receives a sideways or lateral kick, at first in one direction transverse to the direction of propagation. But as the compensating effect of the other mass catches up, the push from the kick fades out, only to be replaced by a growing sideways kick in the opposite direction—growing because the pulse from the nearer source is no longer present to counteract the transverse push caused by the more remote mass.

The strength of the kick from Alpha is governed not by the mass of Alpha alone, but by its mass times its change in velocity, or more precisely, the change in the momentum of Alpha. Likewise for mass Beta. And the law of conservation of momentum tells us that any change in the momentum of objects, big or little, must exactly cancel. But what upholds this law of conservation of momentum? The grip of spacetime on mass. That grip manifests itself in these pulses rolling out through space, first a pulse one way as spacetime staggers under the blow of the one mass, then the other pulse, spacetime's reaction to the counterblow of the other mass. That's a poor man's picture of a gravity wave!

Shudder though it does, spacetime manages to uphold the law of conservation of momentum for every known event on Earth and almost every impact out among the stars. Not for every event, however! A mass shaped oddly enough that implodes asymmetrically enough into a bind-

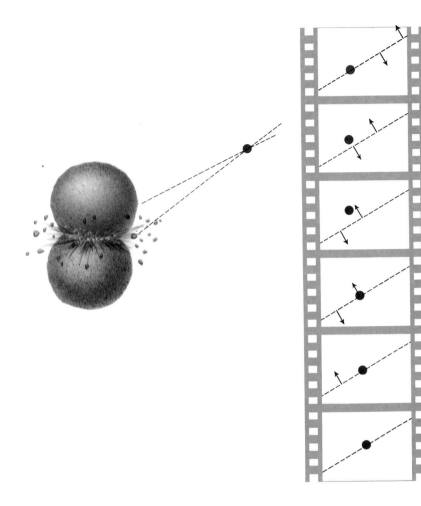

Generation of a gravity wave (left) and its influence on a distant test mass (film strip at right). Frame 1 (bottom) shows the location of the test mass (red) relative to the surrounding space before the disturbance arrives from either of the two colliding spheres. Frame 2 (just above) shows both the test mass, still undisturbed, and—symbolically (black arrow)—the transverse gravity-wave pulse generated by the sudden stoppage of the nearer iron sphere. Frame 3: That pulse hits the test mass. Arriving from the lower left is the oppositely directed arrow symbolizing the gravity-wave pulse generated by the sudden stoppage of the other, more remote, sphere. Frame 4: The test mass, set into motion by the first pulse, has now moved a bit away from its original location relative to the surrounding space. Frame 5: The second pulse hits the test mass. Frame 6: The gravity wave has gone by.

ing tight enough will give spacetime a one-sided kick and in consequence receive itself an oppositely directed kick. Perhaps exactly by this mechanism, Jacob Bekenstein points out, a long known and puzzling neutron-star fragment acquired its unexpected velocity. This fleeting glimpse of an abnormal gravity wave clears the stage for the normal gravity wave to reappear and display itself more fully, now no longer dressed in Newtonian garb.

We throw away, bit by bit, the Newtonian clothes. Before the undressing, we see the gravity wave derived from an outmoded but nevertheless useful idea: gravity, like electromagnetism, is a force that propagates through a flat spacetime background. In contrast, Einstein strips gravity to its basics and displays it to us as manifestation of spacetime curvature. No free-float test mass subject to a passing gravity wave ever experiences any kick, lateral or otherwise. Free float is free float is free float! Gravity manifests itself locally, not in any supposed acceleration of one free-float test mass relative to Newton's fictitious absolute space, but in the relative acceleration of two nearby test masses. When we accept this principle of free-float, we do not throw away the concept of "lateral kick," but translate it into clearer language and observationally testable terms. Acceleration of two nearby test masses, the one relative to the other, is—we know by now—only another way to speak of spacetime curvature. A gravity wave manifests itself through a traveling disturbance in spacetime curvature as definitely as a water wave displays itself through a traveling ripple in the pristine planarity of a pond. But whatever insights we gather from quasi-Newtonian reasoning about the details of that gravity wave have to be checked against—and corrected where necessary to accord with—the great governing principle of geometrodynamics: in any region where space is empty, spacetime curvatures have to balance. In other words, everywhere in the emptiness outside the source of the radiation every little cube of spacetime has to display zero moment of rotation.

Let's jump to the conclusions, foregoing both the mathematics of the exact analysis and the details of the quicker route to the answer via quasi-Newtonian reasoning plus corrections. Let's examine the relative acceleration of two test masses to explore what happens when a pulse of gravitational radiation sweeps by. Prior to the arrival of the wave, they experience negligible acceleration; we've placed them too far away from the source for its static gravitational field to have any significant effect. Moreover, when the test particles' line of separation lies in the direction of travel of the wave, it imparts to them no sudden change in separation. The separation does change, however, when the line of separation is transverse to the direction of propagation. The separation suddenly increases as the front of the pulse passes over the pair of particles, provided that the front of the pulse has negative spacetime curvature. A sudden decrease in separation takes place as the trailing positive-curvature part of the pulse goes by. Or the other way around: first the decrease then the increase. Which happens first depends on the angle between two reference directions. One reference direction is the line of the head-on encounter between Alpha and Beta. The other is the direction to the locale where our two test masses float. Those midget masses experience no change at all in their separation when the angle between the two reference directions is 0 degrees or 180 degrees (observation along the line of impact) or 90 degrees (observation at right angles to the line of impact). In other words, if Alpha plunges down a boomerang shaft straight from the North Pole and at Earth's center smashes into a like mass Beta zooming straight up from the South Pole, observers on Earth's surface will find the resulting pulse of gravitational radiation strongest in the temperate latitudes of the northern and southern hemispheres but zero at the poles and zero at the equator. One final feature about the response of the two test masses is important. Does their line of separation lie in the plane of the two reference directions? Or at right angles to that plane? In the two cases the accelerations are equal in magnitude but opposite in sign. Apart from these interesting features about directionality, what can we say about the strength of a gravity wave and how it compares and contrasts with the strength of an electromagnetic wave?

For waves of both kinds the motive force falls off only as the inverse first power of the distance from the source—not the faster inverse square fall off of the field of a static electric charge of stationary center of mass. In both cases the wave field is amplified as compared to the static field because the preexisting field, kinked by confinement to a conduit of fixed thickness but ever-growing radius, is stretched to an ever-growing extent.

There is this one great difference between waves of the two kinds: how they depend on what happens in the source. This contrast appears most clearly when we turn from one charge, or two masses, to the wiggle-waggle of a whole glob of charge or mass. One charge and its acceleration decide the strength of the electromagnetic wave:

$$\text{electric field} = \begin{pmatrix} \text{angle-} \\ \text{dependent} \\ \text{factor} \end{pmatrix} \times \frac{\begin{pmatrix} \text{electric} \\ \text{charge} \end{pmatrix} \times \begin{pmatrix} \text{its sudden} \\ \text{change in velocity} \end{pmatrix}}{\begin{pmatrix} \text{distance} \\ \text{to locale of} \\ \text{observation} \end{pmatrix} \times \begin{pmatrix} \text{time taken for} \\ \text{sudden change} \end{pmatrix}}$$

To restate this finding most illuminatingly, and so that it will apply not only to a single charge but to a whole distribution of electric charge, we focus attention on the so-called *electric moment* of this charge distribution: charge times displacement from some chosen point of reference in the case of a single charge; the sum of such products when there is more than one charge. An additional item of terminology also proves useful, *time rate of change,* and different orders of time rate of change. It prepares the way for this new and useful language to speak of velocity as "first time rate of change of displacement." Translated into this new terminology, what matters for the radiation from a single charge—its change in velocity divided by the time required for that change in velocity—transcribes itself into "second time rate of change of displacement." Better formulated, what counts is not this second time rate of change of displacement by itself, but this quantity multiplied by the strength of the electric charge—that is, the second time rate of change of the electric moment. And this conclusion applies as well to the electric field produced by a whole writhing array of many electric charges as to the sudden jolt of a single charge:

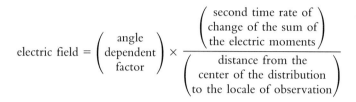

$$\text{electric field} = \begin{pmatrix} \text{angle} \\ \text{dependent} \\ \text{factor} \end{pmatrix} \times \frac{\begin{pmatrix} \text{second time rate of} \\ \text{change of the sum of} \\ \text{the electric moments} \end{pmatrix}}{\begin{pmatrix} \text{distance from the} \\ \text{center of the distribution} \\ \text{to the locale of observation} \end{pmatrix}}$$

As the strength of an electric wave is proportional to the onesidedness of a sudden jolt in a distribution of charge, so the strength of a gravity wave is proportional to the *sudden* change in the nonsphericity of a distribution of mass. No nonsphericity, no radiation! No matter how massive an aging star may be, when at last it collapses under the unrelenting inward pull of gravity, it generates no gravity wave when its distribution of mass is spherically symmetric to begin with and remains spherically symmetric all the way into final crunch. Nonsphericity, so essential for radiation, is a many-splendored thing. Every lump and bump contributes its mite to a measure of nonsphericity commonly called *reduced mass-quadrupole moment.*

Five numbers mathematically express this measure of nonsphericity. I like to watch a star-crunch analyst figuring a sample one of these five numbers. On the grid of his chart he divides up the whole mass into tiny cubelike chunks. For each he figures its mass. Then the location of the center of mass for the entire object. Then through that center he draws three perpendicular lines, which he calls east, north, and up. With these directions as guides, he measures off for each bit of mass how far it lies to the east of the center, and how far to the north. He multiplies those two distances—each in meters—by each other and by the amount of mass in that bit—also in meters. These products he adds up for each bit of mass to get selected numbers that measure the reduced quadrupole moment. He calls this quantity the north-by-east component of the quadrupole moment. The four components corresponding to other combinations of directions make up the other four numbers that give the reduced mass-quadrupole moment. In terms of this measure of nonsphericity, the strength of the gravity wave takes on the form

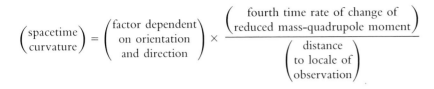

$$\begin{pmatrix} \text{spacetime} \\ \text{curvature} \end{pmatrix} = \begin{pmatrix} \text{factor dependent} \\ \text{on orientation} \\ \text{and direction} \end{pmatrix} \times \frac{\begin{pmatrix} \text{fourth time rate of change of} \\ \text{reduced mass-quadrupole moment} \end{pmatrix}}{\begin{pmatrix} \text{distance} \\ \text{to locale of} \\ \text{observation} \end{pmatrix}}$$

This equation reveals what it takes to produce a gravity wave. A mass asymmetry that's static? No gravity wave! A mass asymmetry that undergoes a steady change? Then, too, there's no gravity wave. What about two faraway masses that stay on steady course? They give rise to a mass-quadrupole moment that changes with time in a complicated way. However, they do not shock the fabric of spacetime, nor do they set it aquiver. They cause no radiation. To do that takes a course that alters so abruptly as to produce no mere steady change in the rate of change of quadrupole moment, nor even any steady change in the rate of change of rate of change of this asymmetry, not even any steady rate of change of rate of change of a rate of change of quadrupole moment. Radiation demands more, a fourth time rate of change of reduced quadrupole moment: that is, a steady rate of change of rate of change of rate of change of rate of change of mass nonsphericity!

A close look at the body of a powerful bird discloses its bulging wing muscles. They propel the rise and fall of the widespread wings. The

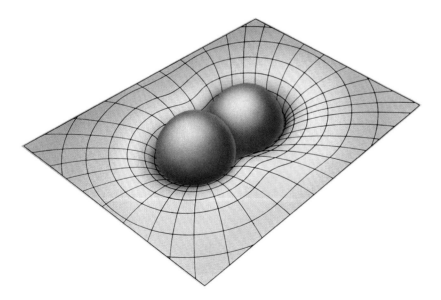

The departure of mass from sphericity is measured by "reduced quadrupole moment." It shows up in the departure of space geometry itself, nearby, from the exact sphericity of standard Schwarzschild geometry.

muscle that drives the gravity wave rippling out through space displays itself equally clearly when we examine the gravity in the immediate surroundings of the radiating mass. The spacetime curvature there—what we otherwise think of as the tide-driving acceleration—departs from the simple dependence on distance reflected in the expression (mass)/(distance from the center of mass)^3, (Chapter 5). To represent the observed deviations from this simple standard law of influence caused by lumps, bumps, bulges, and other departures of the mass from sphericity, the measurer of gravity finds she has to add an extra term to this simple standard formula for the tide-producing curvature. This correction term is of the form (orientation- and angle-dependent factor) × (reduced mass-quadrupole moment)/(distance)^5. Merely to display such an asymmetry in its distribution of mass does not, however, qualify an object to radiate a gravity wave. Neither can a bird fly merely by flexing his wing muscles. He must alternately flex and relax them to make his wings drive him through the air. Likewise the mass asymmetry must change rapidly if the source is to radiate a powerful pulse of gravitational radiation. That's a rough and ready way to become comfortable with the fact that the decisive measure of source strength is not the quadrupole moment itself, but its fourth time rate of change!

The Detection of Gravity Waves

One sudden asymmetric change in the pattern of mass-in-motion gives one pulse. A whole string of sudden changes produces a whole string of pulses, a train of waves, a pattern of sphere-shaped zones of positive and

Left: *Periodic gravity waves arise from two stars that swing each other around in orbit, as nearly half of all stars do. The time to go around varies from years to hours, minutes, or even fractions of a second. Several times a year, if today's expectations are correct, two objects as compact as a neutron star or black hole spiral in to catastrophic encounter and produce on Earth an alteration of detector dimensions as great as one part in 10^{22} or more.* Right: *A star, endowed through normal rotation with an equatorial bulge, has grown too old and too massive. It collapses and its core forms a rotating neutron star or black hole. Such an event generates a burst of gravity waves with a wiggle-waggle time of the order of a millisecond. Perhaps once every 30 years or so, the collapse of a star produces gravity waves powerful enough to jiggle detector dimensions of the order of one part in 10^{18}.*

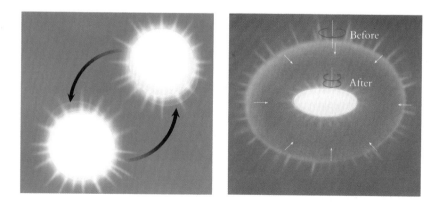

Chapter Eleven

negative spacetime curvature rippling out through space. The disturbance, pulse, or wave train makes itself known by altering the separation between a pair of test masses disposed transverse to the local direction of travel of the wave. To detect this separation is the Jason's Fleece sought by a dozen gifted experimentalists dotted here and there about the globe. How do we translate into the terms they care about—jolt in *separation*—the thing that spacetime cares about—jolt in *curvature*?

Two test masses are happily climbing up through spacetime on parallel or nearly parallel worldlines before the first gravity wave ripples over them. When the positive-curvature region of the first gravity wave reaches them, its action is immediate. It starts bending the two test masses toward each other. They continue to bend toward each other during the time of passage of this first positive-curvature disturbance of spacetime. That is, they accumulate, as did the boomerangers of Chapter 4, a speed of approach given, relative to the speed of light, by the product of three factors: the separation itself, the average curvature, and the time of action of this curvature. At this point we are happy to recall that the curvature which is rippling by is proportional to the *fourth* time rate of change of the mass asymmetry of the source. The bend it imparts, accumulated over time, is proportional to the *third* time rate of change of this mass quadrupole. And the decrease in the separation itself is determined by the average speed of approach multiplied by the time, that is, by the accumulation of velocity over time. And this quantity is proportional in its turn to the *second* time rate of change of the mass asymmetry. This finding displays remarkable simplicity when translated from the separation itself to the fractional change in separation, and that is the quantity

From left to right: *A gravity wave encounters a black hole and sets it into vibration. Ten time units later part of the original wave has gone down the hole (region below the white line), but an additional new wave is coming off because the black hole was distorted from sphericity by the impact of the original wave. After twenty and thirty time units, some of the newly generated waves have propagated down the hole while others are rippling out to the faraway. Raymond Idaszak and Donna Cox did the graphic rendering and animation for these diagrams from numerical simulation carried out on a supercomputer at the University of Illinois by David Bernstein, David Hobill, and Larry Smarr.*

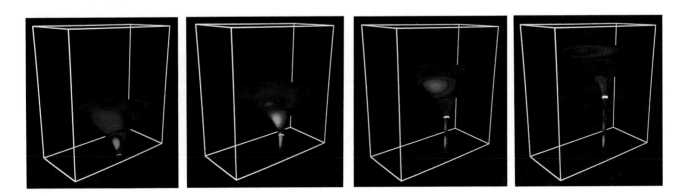

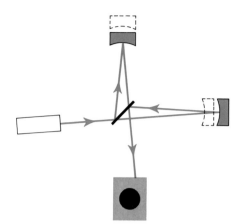

The proposed MIT-Caltech gravity wave detector will (1) take the beam from a laser (left), (2) split it by a device (center) analogous to a half-silvered mirror, (3) send the one half-strength beam to one faraway mirror (top) and the other to the other faraway mirror (right), (5) allow these beams to undergo many many reflections (not shown), and (6) recombine them at the detector (bottom). A gravity wave incident on the Earth will slightly shorten the 4-kilometer distance to the one mirror and slightly lengthen the 4-kilometer distance to the other mirror. This relative alteration in the path length of the laser beams, if big enough, amplified enough, and picked up by detectors sensitive enough, will reveal the passage of the gravity wave.

of number one interest to every experimentalist designing a device to detect a gravity wave:

$$\begin{pmatrix} \text{fractional shrink-} \\ \text{age of detector} \\ \text{dimension} \end{pmatrix} = \begin{pmatrix} \text{factor dependent} \\ \text{on orientation} \\ \text{and direction} \end{pmatrix} \times \frac{\begin{pmatrix} \text{second time rate of} \\ \text{change of reduced mass-} \\ \text{quadrupole moment} \end{pmatrix}}{\begin{pmatrix} \text{distance} \\ \text{to locale of} \\ \text{observation} \end{pmatrix}}$$

Not much! That's our first impression when we look at the fractional shrinkage in detector dimension to be expected—according to this formula—from the selected astrophysical events depicted in the illustrations on pages 202 and 203.

I like to compare these visions of the achievable with the prospectus that Columbus had to put before Ferdinand and Isabella before they would grant him the wherewithal. In our time, gifted experimenters decade by decade devise ever more ingenious ways to detect ever smaller fractional changes in detector dimension. Starting with Joseph Weber's one-ton aluminum bar detectors of the early 1960s, physics had come by the end of 1989 to the point where groups at the Massachusetts Institute of Technology and the California Institute of Technology were working together on 1.5-meter and 40-meter prototypes of an intended 4-kilome-

Prototype gravity-wave detector, California Institute of Technology, Pasadena. The laser beam is tailored (lower right) for entry into the beam splitter (located where the two long light pipes meet, just to the left of center in the photograph). The mirrors at the ends of these two evacuated light pipes lie outside the boundary of the photograph.

Chapter Eleven

ter interferometric detector. They and other enterprising experimenters have embarked on a Columbus-like voyage of discovery. What they will find may diverge by a like amount from expectation. Once again their discoveries will count in direct proportion to the surprise that they occasion. Once again all mankind will be the gainer.

In the meantime we already have an exciting indirect confirmation that gravity waves exist—not through their action on any receptor, but through the energy they carry away from a whirling pair of neutron stars. That particular binary pulsar first revealed itself to Joseph H. Taylor, Jr. and Russell A. Hulse by periodic pulses of radio waves picked up on the huge dishlike antenna at Arecibo in Puerto Rico. As one of these neutron stars spins on its axis, its magnetic field spins with it, giving timing comparable in accuracy to the best atomic clock ever built. Thanks to this happy circumstance, Taylor and his colleagues have been able to follow the evershortening separation of the two stars and the ever higher speed they attain as they slowly spiral in toward an ultimate catastrophe some 400 million years from now. The timing of the orbits gives us a measure of energy lost as the stars spiral in. No reasonable way has ever been found to account for the thus observed loss of energy except gravitational radiation. As of September 1989, 14 years after first observation, this loss of energy agrees with the rate predicted by theory to better than one percent.

In a gravity-wave experiment in Pisa, Italy, each of two 400-kilogram mirrors is isolated from highway and other disturbances by a three-dimensional seismic noise seeker-attenuator. It consists of a cascade of seven vertical gas springs (not all visible in the photograph), each weighing 100 kilograms, interconnected by vertical wires seven-tenths of a meter long.

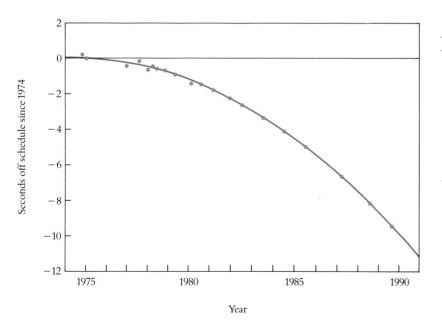

The two whirling neutron stars furnish a giant clock, whose time-keeping hand is the ever-turning line that separates the centers of those two stars. That hand does not today keep the "slow" schedule (straight line) one might have expected from its timing as measured in 1974. The downward sloping curve shows gravity-wave theory's prediction of the shortening in the time required to accumulate any specified number of revolutions. The circles show the actual observed shortening in that time.

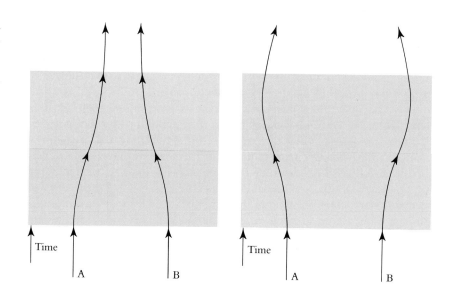

The passage of a pulse of gravitational radiation leaves test masses A and B on approaching worldlines even though originally they were marching forward through spacetime on parallel worldlines. Left: The positive curvature region of the pulse (pink) is first to roll over the test masses. Right: The negative curvature region (green) arrives first. In both cases approach results.

How does a gravity wave, a ripple in a spacetime everywhere totally void of mass–energy, nevertheless manage to transport energy? Because it can deliver energy to mass! And, far far away, it ultimately does deliver energy to mass. Thus, two worldlines A and B, originally parallel, bend together under the influence of the positive spacetime curvature in the first part of the arriving pulse (indicated in pink in the diagram on this page). When the second part of the pulse rolls over them, of equal negative spacetime curvature (green), they would be bent back through an equal "angle" (relative velocity) and end up nonapproaching if bending depended solely on curvature. However, bending depends not only on curvature but also on separation, and the separation at this point had already been diminished compared to its original amount. Therefore not all of the original bending is undone. After the pulse has gone by, the two test masses end up with a net velocity of approach. Moreover, out of this velocity of approach useful energy can be extracted. That's one way, and a very direct way, to recognize that a gravity wave transports energy.

Gravity waves and pulses of gravity radiation are sweeping over us all the time from sources out in space of many different kinds. At detecting them, however, we are no better than the primitive jungle dweller, unable to detect and even totally unaware of the radio waves that carry past him every minute of the day music, words, and messages. However, experimentalists of today are working out ingenous technology and building detector instrumentation of ever-growing sensitivity. Few among them have any doubt of their ability to detect pulses of gravity radiation from one or another star catastrophe by sometime in the first decade of the new century.

Astronomy uses signals of many kinds—light, radio waves, and x-rays among them—to reveal the secrets of the stars. Of all signals from a star, none comes out from deeper in the interior than a gravity wave. Among all violent events deeply to be probed by a gravity wave, none is more fascinating than total collapse of a mass into . . . a black hole!

H.K.WIMMER

12

Black Holes

Oh Reader, think on me.
Think on the mystery summed up in me,
Ask how many more black holes like me
Float today across the heavens.
Gape at the grip of my gravity on each new wisp of gas
That comes too close.
I crunch gas and radiation alike.
Even the space that surrounds them
I crush out of existence.
Am I your predictive model
For the fate of the universe itself?

A painter's vision of a black hole accreting gas from a nearby star (lower right).

A black hole differs dramatically from a star of any other kind. Other stars contain both matter and mass. In contrast, the black hole is disembodied mass, mass without matter. The Cheshire cat in *Alice in Wonderland* faded away leaving behind only its grin. A star likewise fades away when it falls into an already existing black hole, or when it collapses to make a new black hole. All traces of the star disappear—its matter, its sunspots, its solar prominences. Only gravitational attraction remains behind—the attraction of disembodied mass. This attraction continues to hold in orbit any planet that was in orbit around the star while it lived.

The Italian physicist Tullio Regge once suggested to me that *Nimmersatt,* never satisfied, might have been a better term than black hole to indicate a totally collapsed star: the bigger it is, the more it eats; and the more it eats, the bigger it grows.

At the center of a black hole is the point of *crunch*. There the matter that once composed the star is crushed out of existence. In that crunch matter disappears, with all its particles, pressures, and properties. Pure matter-free mass remains. Any additional gas, planet, or star that falls in suffers the same fate, crunch. The black hole deals out that fate, not by lottery, but by law. Matter that transgresses the perimeter of no return is caught and crunched. Matter that stays outside the perimeter is spared.

The perimeter of no return is commonly referred to as the *horizon* of a black hole. A traveler approaching in a spaceship feels no bump, no acceleration, no jar, as he crosses this unmarked boundary. Once inside it, however, he cannot escape, no matter how powerful his rocket engine. Nor can any light signal or radio message get out from him to the outside world. They, like him and his ship, contribute nothing but blackness to a faraway observer. Within a short and firmly fixed time, the signals and the traveler are obliterated at the central *singularity,* the point of crunch.

The first person known to have considered seriously an object so massive or so compact, or both, that light could not escape from it was the Reverend John Michell. This amateur British astronomer devised the experiment carried through by Henry Cavendish to determine the mass and density of the Earth, the results of which Cavendish published in 1798. In 1783 Michell sent to Cavendish a calculation, based on Newtonian gravity, that showed that a sphere of the same density as the Sun but five hundred times larger would have an attraction so powerful that "all light emitted from such a body would be made to return towards it." Pierre-Simon Laplace independently came to the same conclusion in 1795 and went on to reason that "it is therefore possible that the greatest luminous bodies in the universe are on this very account invisible."

Michell had considered an object with the density of the Sun, which equals the density of water, while Laplace had considered a star with the same density as the Earth, which is five and a half times the density of water. Such a uniformity and modesty of density, we realize today, is totally incompatible with a gravitational influence powerful enough to hold in light. But how does matter respond to such a squeeze? The answer came from modern quantum theory combined with new knowledge about the structure of matter. Occasional theorists continued to make calculations about the properties of the objects that we now call black holes. The advent of the term *black hole* in 1967 was terminologically trivial but psychologically powerful. After the name was introduced, more and more astronomers and astrophysicists came to appreciate that black holes might not be a figment of the imagination but astronomical objects worth spending time and money to seek.

Today we have reasonably good evidence for the existence of a black hole or more. Nothing did more to open the door to the first sight of these objects than the advent of x-ray astronomy. In December 1971 the Uhuru orbiting x-ray observatory detected x-ray pulsations from the object Cygnus X-1 (code name for the first x-ray emitter in the Cygnus region). The x-rays varied substantially in intensity on timescales from 100 milliseconds to tens of seconds. A search by a radio telescope of the region of the sky close to the source led in the same year to the discovery of a source of radio waves. By 1972 it was clear that the radio waves had to be identified with an object seen by optical telescopes. This visible star was already known to move in orbit about an invisible companion with a

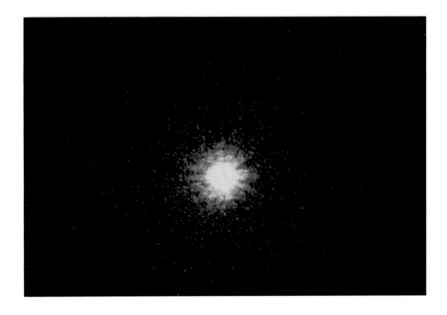

X-rays from the first discovered black hole, the one now known as Cygnus X-1.

Black holes for which there was substantial evidence as of September 1989

Astronomical designation of black hole	Mass (in solar masses)
Cygnus X-1	9.5
LMC X-1	2.6
AO 620-00	3.2
LMC X-3	7.0
SS 433	4.3
Black hole at center of galaxy	3.5×10^6

Uncertainties in the mass are of the order of 20 to 50 percent.

5.6-day time of revolution and to have a mass in excess of 10 solar masses. The two stars are like the participants in a formal dance with lights turned low. Only the white dresses of the girls are visible as they are whirled around in the arms of their black-suited companions. From the speed of the girl and the size of the circle in which she swirls, we know something of the mass of the invisible companion. By similar reasoning, it was possible to conclude by 1972 that the optically invisible x-ray emitter has a mass of the order of 9.5 solar masses.

Why does the mysterious object give off x-rays? And why does the intensity of the x-rays vary rapidly from instant to instant? The gas wind from the visible companion varies from instant to instant like the smoke from a factory chimney. This gas, falling on a compact object, gets squeezed. To picture how and why of the squeeze, look from a low-flying plane at the streams of automobiles converging from many directions on a football stadium for a Saturday afternoon game. The particles and the gas are pushed together as surely as the cars in the traffic. The compression of the traffic raises the temper of the driver, and the compression of the gas raises its temperature as air is heated when pumped in a bicycle pump. However, because the gas falls from an object of millions of kilometers in size to one a few kilometers across, the compression is so stupendous that the temperature rises far above any normal star temperature, and x-rays come off.

The timescale of the fluctuations in x-ray intensity depends on the size of the object that is picking up the star smoke, a size less by a fantastic factor than that of any normal star. Could the object be a white dwarf? No, because such a star would be visible. A neutron star? No, because even matter compressed so tightly that it is transformed to neutrons cannot support itself against gravity if it has a mass much over two solar masses. No escape had been found from concluding that Cygnus X-1 is a black hole. This great discovery transformed black holes from pencil-and-paper objects into a lively and ever-growing part of modern astrophysics.

In addition to the probable black holes listed in the table stand the most distant lighthouses—and most mightly powerhouses—of the universe, quasistellar sources or quasars, which number on the order of a million. No acceptable explanation has ever been put forward for their immense energy output except accretion onto a rotating black hole of mass in the range of 10^8 to 10^{12} solar masses.

Much attention has gone in the 1980s to a presumptive black hole with a mass about three and a half million times the solar mass and a Schwarzschild radius of about ten million kilometers. It floats at the center of our galaxy, the Milky Way. Around it buzz visible stars of the everyday kind, most of them fated to fall eventually into that black hole and increase its mass and size. That stars close to the center of our galaxy go around as fast as they do is one of the best indicators we have for the presence, and one of the best measures we have for the mass, of the central black hole, which is itself invisible.

The Penrose Process

To tell their students something both new and true, something that will grip them with its power and surprise, is the time-honored obligation of teacher-researchers. Roger Penrose faced this challenge at the University of London in the 1960s and 1970s. Riding in on the London-bound train to his afternoon lecture in 1969 and searching for something new and true to report on the black hole, he suddenly hit on what has ever since been called by everybody else the *Penrose process*. It is a scheme to extract energy out of a spinning black hole, otherwise known as a live black hole, which is formed by the collapse of a rotating star or by the accretion of matter that was in orbit around an already existing black hole. The discovery of this mechanism proved to be the doorway to a whole series of findings in black-hole physics, the work of many hands.

The Penrose process is a trade of energy that can be carried out between two masses or two "particles" in the immediate vicinity of a rotating black hole, in a region called its *ergosphere*. In this trade one particle, A, gets energy; the other particle, B, gives up energy. That trade is similar to a transaction in which friend B gives $100 to friend A. But how does B manage to make a $100 gift when, to begin with, he has only $20 in all the world? He hands to A the $20 plus an IOU for $80.

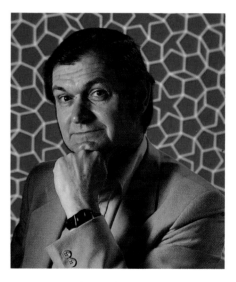

Roger Penrose

Born August 8, 1931, Colchester, England. Penrose, a professor of mathematics at Oxford University, proved in 1965 that under certain very general circumstances spacetime will inescapably undergo crunch. He continues to be a source of deep insights on spinors, twistors, and geometrodynamics generally.

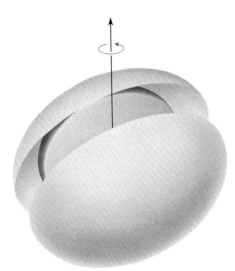

The critical domains of a rotating black hole. The boundary of the inner sphere (gray) marks the horizon—from inside it no object can escape, even one traveling at the speed of light. Between the horizon and the outer ellipsoid (green), or "static limit," lies the ergosphere. There a particle can run up an energy indebtedness so great that its capture by the black hole, far from augmenting the mass of the black hole, diminishes it.

After this transaction B goes around with less than nothing—with the debt of $80 affixed to him. The Penrose process of energy exchange works the same way. Particle A gains 100 units of energy from its interaction with particle B. If particle B had only 20 units to begin with, then particle B is 80 units in the hole after the interaction. Shortly afterward particle B, encumbered with its 80-unit debt, orbits across the horizon, the inner surface of the ergosphere, and is irrevocably captured by the black hole. The black hole takes over its debt and ends up with a little less energy of rotation than it had before and a little less speed of spin. Through the intermediation of B, A has robbed the black hole of energy, a theft that A could not have pulled off directly without itself undergoing capture and destruction.

Only in a special place is some object able to incur such a debt of energy that its energy becomes negative. That place is the ergosphere. Any particle put into the appropriate state of motion (right speed, right direction) in the ergosphere will go negative in energy—just as any individual under appropriate circumstances in a modern debt-recognizing society will go negative in net financial worth.

Debt-recognizing societies and ergospheres comprise only a small fraction of the universe. At no place is an ergosphere known or expected except in a doughnut-shaped region around a rotating black hole. This region is bounded on the inside by the horizon, the perimeter of no return, the crossing of which by a particle makes inevitable the ultimate crunch of that particle. The ergosphere is bounded on the outside by a second closed surface, the so-called *static limit*. Both surfaces have an egglike deformation unless the black hole is not rotating at all. In that special case both surfaces are spherical and coincide, so that the ergosphere between them is pinched out of existence.

Outside the static limit, a space traveler can maintain her spaceship in a "static" position above the black hole from which she will always see the same configuration of stars, as the black hole rotates beneath her. The only requirement is that the blast of her downward-directed spaceship engines be sufficiently powerful. The static limit is reached when no conceivable rocket engine is able to preserve for the observer a forever static view of the faraway. If a space traveler decides to risk even more extreme conditions and support himself inside the static limit, at the cost of turning up his rocket engine to a yet more powerful blast, he can still maintain his new location relative to the black hole, so many kilometers above the horizon but now so many kilometers below the static limit. However, he is willy-nilly dragged around in the direction of the rotation of the black hole. He can arrange to go around in this rocket-supported orbit at a faster rate or slower rate, but not at a zero rate. To maintain a fixed distance from a black hole, to keep some specified dis-

Chapter Twelve

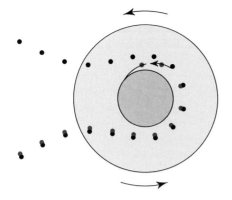

The Penrose process extracts energy from a rotating black hole. The energy-greedy particle A (black) brings along into the ergosphere the sacrificial victim B (red), held in tight embrace. There A employs part of B's mass to create an explosive charge that drives A forward with great velocity out of the ergosphere. Recoiling, B still goes forward, but with much reduced velocity, and falls into the black hole. A ends up far away with a net energy increase in excess of the original total energy of B. The mass of the black hole decreases correspondingly, despite the fact that it has gobbled up B.

tance above the horizon, and thus to withstand the pull of gravity demands a sufficiently powerful and downward-directed rocket engine but presents no insuperable difficulty.

At a specified distance above the horizon of a nonrotating black hole, the energy of a planet, boulder, or particle is least when it is hovering motionless, sustained by a rocket engine of the appropriate strength. At the truly risky position just above the horizon, the energy of the hovering particle is practically zero. That is, the work it takes to drag that particle from where it is to great distances, expressed as an energy, adds up to almost all the energy that the particle has, at rest, very far from the black hole. In this sense we ourselves, through the work we do in dragging the particle out, are responsible for 99 percent or more of its mass.

The energy of the particle located at some specified distance from the center, one would surely say, is least when it is not circling at all, merely hovering. This conclusion, so natural and so true for a nonrotating black hole, is wrong for one that is rotating. The rotation of the black hole alters conditions in the space around it. The particle must slowly circle the rapidly turning black hole in the direction opposite to the direction of rotation of the black hole if it is to have zero energy of motion. For a particle in the ergosphere of a rotating black hole, zero energy of motion implies nonzero motion, and zero motion implies nonzero energy of motion. A particle can go around the black hole in a rocket-assisted circular orbit opposite to the black hole's direction of rotation fast enough to have a negative energy. In other words, to drag it from its present state of motion to a condition of rest at infinite distance from the black hole costs more in energy than the total energy of the particle as thus removed. This is how it comes about that a perfectly natural particle moving in a perfectly natural direction around a perfectly natural rotating black hole can qualify as the debtor particle B of the Penrose process.

Naturally no particle goes around with a rocket engine prepared to sustain it in an unnatural circular orbit. With no such assistance, particle B, starts off tangent to the circle, departs more and more from that safe track through the ergosphere, falls through the horizon into the black hole, and is crunched out of existence. Then how can particle A, coming in from a great distance. ever expect to find in the ergosphere any ill-fated particle B from which it may hope to borrow energy? It can't.

With no potential sacrificial victim waiting in the ergosphere to be exploited, A adopts a different policy. It brings B with it on its inward journey from afar. They could continue that companionship through the ergosphere and off again to great distances. Because A recognizes there is no profit in energy to be made that way, it ditches B in the ergosphere. For this purpose, A carries in with it from great distances an explosive charge with which to disengage B. As this charge explodes inside the ergosphere, it knocks B out of its momentarily forward direction of motion around the black hole into a retrograde motion, a condition of negative energy, a debtor state. A, by contrast, has its already forward motion still further increased and escapes to great distances with far more energy than the push of the explosive, normally employed, could ever have given it. The source of this extra energy? The energy that B had to begin with plus the indebtedness taken on by B—a debt ultimately taken up by the rotating black hole itself. It does not rotate as fast as before, in consequence of its encounter with A.

If Penrose thus discovered a tool for interacting with a rotating black hole, Demetrios Christodoulou pushed this tool to the limit. He explored the whole range of possibilities for the initial directions of A and B, their relative masses, and the energy of explosion that drives them apart. He focused his attention not on particle A and its gain but on the black hole and its loss. For this purpose he capitalized on the fact that energy is conserved. Whatever the particle or particles gain in mass-energy the black hole loses. A similar law of conservation applies to angular momentum, that is, the product of the linear momentum of outgoing particle A multiplied by its distance from the center. Whatever angular momentum is carried off by particle A, the black hole has lost. If enough angular momentum is extracted from the black hole, then it will no longer rotate. Its ergosphere will collapse out of existence. At this point, the Penrose process will come to the end of the road.

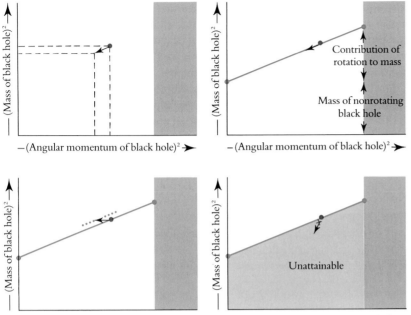

Upper left: *Energy and angular momentum of a rotating black hole before (top and right dashed lines) and after (bottom and left dashed lines) they are changed by the Penrose process.* Upper right: *There is a most efficient way to extract energy from a black hole; it puts the energy and angular momentum of the black hole after the process on the same Christodoulou line (green) where it was before the extraction was performed. Different points on this line correspond to different masses for the black hole and different speeds of rotation, but identical horizon surface areas. The orange region at the right is unattainable because there the rotational speed of the horizon would exceed the speed of light.* Lower left: *It's easy to conduct the Penrose energy-extraction process less than efficiently, but then the black hole ends up above the Christodoulou line.* Lower right: *There's no way to push the black hole below the line. This discovery provided the first hint that a black hole has entropy.*

Different sequences of Penrose interactions, Christodoulou discovered, yield different payoffs in energy captured from the original rotating black hole. Some manipulations—the diagram on the top of this page shows—are efficient in capturing energy from the black hole; others are inefficient. He discovered in 1970, however, that ways of trading energy with a black hole fall into two categories, irreversible and reversible. When angular momentum is taken away from a black hole, by the Penrose process, so too is energy taken away from it. Then, by shooting in a particle in an off-center impact, that same amount of angular momentum can be added back to the black hole and with it, of necessity, some energy and mass. Following this sequence, the black hole normally will end up with more energy and mass than it had to begin with: the change in angular momentum has been reversed exactly, but the change in energy has not. Moreover, nothing we can do will reverse this dissipation of external mass into black hole mass, assuming that we have to bring the black hole back to its original angular momentum. In this situation a black hole is like an irreversible thermodynamic engine. The diagram shows how a sequence of interactions with a black hole that brings it back to its original angular momentum also brings it back to more than its original mass. Only when the optimum choice of particle masses,

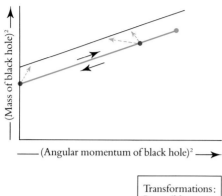

(Mass of black hole)² — vertical axis

— (Angular momentum of black hole)² →

Transformations:
Reversible ——→
Irreversible --->

In a reversible transformation, the black hole stays on the Christodoulou line. An irreversible transformation takes the black hole off the line.

direction of motion, and points of impact is made can reversibility be achieved: return to the original mass as well as the original angular momentum.

During a reversible transformation, the mass and angular momentum of a black hole individually may change, but a remarkable combination of the two properties, what Christodoulou called the *irreducible mass* of the black hole, remains unchanged. In other words, mass and angular momentum before and after reversible change, differ though they do, are nevertheless given by one and the same formula

$$(\text{initial mass})^2 = (\text{irreducible mass})^2 + \frac{(\text{initial angular momentum})^2}{4 \times (\text{irreducible mass})^2}$$

with the same irreducible mass.

Further work by Stephen Hawking in the early 1970s showed that the square of the irreducible mass (expressed in geometric units) is directly proportional to the area of the horizon of the black hole:

$$\text{area of horizon} = 16\pi \times (\text{irreducible mass})^2$$

An apple at rest and the same apple spinning around its axis differ in mass, in energy, and in angular momentum, but they are otherwise the same. With equal justice we can say that two black holes, one rotating and the other not rotating but more massive, are essentially the same when they have the same irreducible mass or the same horizon area.

The Entropy of a Black Hole

A black hole has a fascinating feature: it hides information. A black hole extinguishes all particularities of whatever falls in, and so effectively that the resulting object is characterized by three and only three properties: mass, angular momentum, and electric charge. Measure mass, angular momentum, and charge as precisely as we will, not one hint can we gain from outside as to what went in. Stars? Comets? Dust? Isolated atoms? Quanta of light? Or, better put, what ungodly mixture of all of these?

It takes a stupendous amount of information to describe this mixture in all detail. At this point we may well ask, could any finite number of quantities describe in total detail all that mess of matter and radiation? No would be the immediate answer if the world were described by classical physics. According to classical physics, every physical quantity such as position, or speed, or energy, or angular momentum is described by a number that can take on any one of an infinity of possible values. But

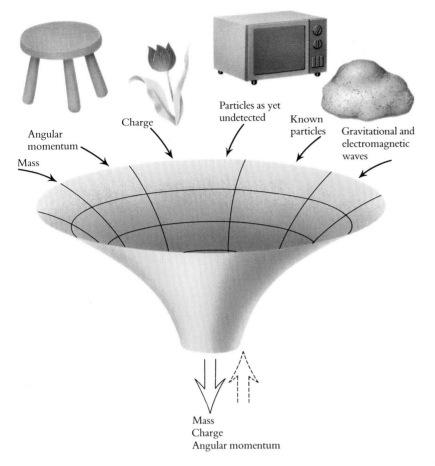

Angular
momentum

Mass

Charge

Particles as yet
undetected

Known
particles

Gravitational and
electromagnetic
waves

Mass
Charge
Angular momentum

No matter what variety of objects fall into a black hole, they leave it possessing at the end—as seen from outside—only mass, angular momentum, and electric charge. Not one hair remains to betray its past.

that isn't true. Angular momentum, for example, can take on only one or another integer multiple of a basic unit of angular momentum. Expressed in conventional units, this basic unit has the value 1.054×10^{-34} kilogram-square-meters per second. In geometric units, where we express both mass and time in meters, this elementary quantum of angular momentum turns out to have the dimensions of an area, a fantastically small area of 2.612×10^{-70} square meters (the square of the so-called Planck length, 1.616×10^{-35} meters). This tiny area is an *elementary quantum of area*. In consequence of this granularity in the very underpinnings of the world itself, the information hidden in a black hole is not infinite, it is finite.

Almost miraculously, the amount of information hidden in a black hole is expressable by a single sharp clear number, the so-called entropy of a black hole. And this entropy, a pure number, the so-called

Bekenstein number, is exactly the surface area of the horizon of the black hole, divided by four times the elementary quantum of area. Brought to light in this way is a marvelous connection between mass in its purest form—black hole mass—and geometry. The exploitation of this connection has become one of the most exciting frontiers in the world of physics.

What is entropy? What is lost information? Human beings first encountered the quantity "entropy" in fundamental investigations by Sadi Carnot (1796–1832) and Lord Kelvin having to do with how much work can be extracted out of a given amount of heat. No engine whatever that takes in a given amount of heat at 1500 degrees Celsius and works by piston expansion or turbine whirl can cool a working substance down to the lowest available temperature, say river water at 20 degrees Celsius, and extract out of that working substance more than a sharply defined amount of heat. No matter that there is still lots of energy still resident in

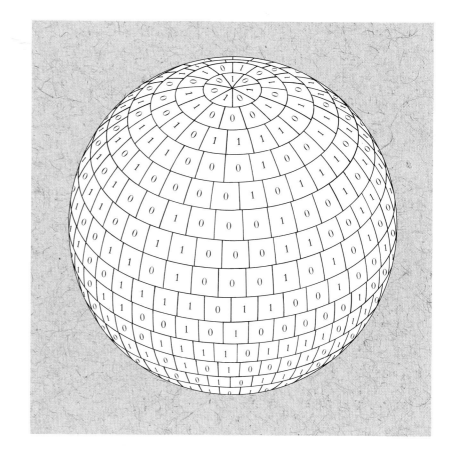

A totally symbolic representation of the Bekenstein number of a black hole. This number counts the number of boxes of Bekenstein size that can be pasted down (in imagination only!) on the horizon of the black hole. Each box contains one yes-or-no binary digit of information about what went in to make that black hole.

Chapter Twelve

the working substance at 20 degrees Celsius in all those rapidly moving molecules. There is no way to get it out! We could get it out if we knew where each molecule is and how fast it is going and in what direction. For then we could put a microscopic target in its path and extract the energy of that molecule and put it to use. By doing likewise for all the other molecules, we could leave the working substance in the end completely devoid of all agitation whatsoever. Then we would reduce it to the absolute zero of temperature. But it would take an enormous amount of information about all these molecules to accomplish this miracle. That information simply isn't available. We have come to realize in this century that entropy represents unavailable information.

It's easy to lose information, impossible to get it back. A liter jar half filled with 100-degrees-Celsius water contains faster moving molecules than a liter jar half filled with 0-degree-Celsius water. We lose this information about which molecules are faster moving and which are slower moving when we mix the two jars. There is a very precise measure of the entropy increase, the loss of information incident on this mixing process. The entropy increase of the combined kilogram of water, expressed in conventional metric heat units, is 12.4 calories per degree Celsius.

We recognize today that the increase of entropy from combining so many hot molecules with so many cold molecules is given by a dimensionless number—in this case, 5.5×10^{24} dimensionless entropy units. This gigantic number measures the number of bits of information that are potentially at our disposal for the operation of a heat engine and that we have now lost because we let hot and cold mix instead of using the temperature difference to do work. The Princeton statistician John Tukey devised the term "bit" of information as an abbreviation for the phrase "binary digit." A bit represents the amount of information contained, for example, in the yes or no answer to the question: Did this particular molecule reside in the hot jar before the mixing?

One afternoon in 1970 Bekenstein—then a graduate student—and I were discussing black-hole physics in my office in Princeton's Jadwin Hall. I told him of the concern I always feel when a hot cup of tea exchanges heat energy with a cold cup of tea. By allowing that transfer of heat I do not alter the energy of the universe, but I do increase its microscopic disorder, its information loss, its entropy. The entropy of the world always increases in an irreversible process like that.

"The consequences of my crime, Jacob, echo down to the end of time," I noted. "But if a black hole swims by, and I drop the teacups into it, I conceal from all the world the evidence of my crime. How remarkable!"

Bekenstein, a man of deep integrity, takes the lawfulness of creation as a matter of the utmost seriousness. Two or three days later he came back with a remarkable idea. "You don't destroy entropy when you drop those teacups into the black hole. The black hole already has entropy, and you only increase it!"

Bekenstein went on to explain that the surface area of a black hole is not only analogous to entropy, it is entropy, and the surface gravity of a black hole (measured, for example, by the downward acceleration of a rock as it crosses the horizon) is not only analogous to temperature, it is temperature. A black hole is not totally cold!

Although black-hole physics up to this time had had nothing to do with the world of the quantum or the particulate character of matter and energy, according to Bekenstein they are connected. Entropy can't be expressed in natural units unless nature itself points to a natural unit of length or area. Nature, however, as seen in classical physics, is indifferent to the size, or the mass, of a black hole. According to classical physics, in no way does a black hole of ten solar masses differ from a black hole of ten million solar masses except through a simple scale factor. Quantum theory, however, provides a natural scale of length, the quantum of area or 2.612×10^{-70} square meters. One unit of entropy, one unit of randomness, one unit of disorder, Bekenstein explained to me, must be associated with a bit of area of this order of magnitude; that is, multiplied by some then unpredictable numerical factor like two or π. This factor later turned out to be four. Thus one unit of entropy is associated with each 1.04×10^{-69} square meters of the horizon of a black hole.

To get an idea of the magnitude of this natural unit of entropy, let's take a black hole of 3 solar masses; since 1 solar mass equals 1.47 kilometers in geometric units, the mass—or irreducible mass—of this object is 4.41×10^3 meters. The radius of its horizon is twice that; and the horizon area itself is 4π times the square of this radius, or 9.8×10^8 square meters. Clearly, when we divide this figure by the natural area associated with a unit of entropy (1.04×10^{-69} square meters), we end up with an absolutely stupendous number of natural units of entropy; even more entropy units are contained in a larger black hole. By similar order-of-magnitude reasoning, Bekenstein was able to estimate that the surface temperature of a black hole of 3 solar masses is less than a millionth of a degree above the absolute zero of temperature, and still colder for still more massive black hole masses.

The Hawking Process

Studying Bekenstein's conclusions, Brandon Carter and Stephen Hawking were incredulous and decided to find a way to disprove the idea that a black hole has entropy and temperature. For any surface to have a temperature implies the possibility of evaporation from that surface. To show that a black hole has no possibility to evaporate a photon, the elementary quantum of luminous energy, from its surface would be enough to prove that that surface is at the absolute zero of temperature, as one had always conceived of a black hole before. Alternately, if black holes could be shown to emit photons, then they must not be totally black, or totally cold. Hawking set out to determine what happens at the surface of a black hole.

Hawking discovered that quantum theory allows a process to happen at the horizon analogous to the Penrose process. As explained earlier, in the Penrose process two already existing particles trade energy in the ergosphere, the only domain where the brief coexistence of the positive-energy particle and the negative-energy particle is possible. Because the ergosphere shrinks to extinction when a black hole is deprived of all spin, the Penrose process applies only to a rotating, or live, black hole. In contrast, the Hawking process takes place at the horizon itself and thus operates as effectively for a nonrotating black hole as for a rotating one. Furthermore, unlike the Penrose process, it involves a pair of newly created particles.

Space—pure, empty, energy-free space—all the time and everywhere experiences so-called quantum fluctuations at a fantastically small scale of time, of the order of 10^{-44} second and less. During these quantum fluctuations, pairs of particles appear for an instant from the emptiness of space—perhaps an electron and an antielectron pair or a proton and an antiproton pair. Particle-antiparticle pairs are in effect all the time and everywhere being created and destroyed. Their destruction is so rapid that the particles never come into evidence at any everyday scale of observation. For this reason, the pairs of particles everywhere abirthing and adying are called *virtual* pairs. Under the conditions at the horizon, a virtual pair becomes a real pair.

In the Hawking process, two newly created particles exchange energy, one acquiring negative energy and the other positive energy. The negative-energy particle flies inward from the horizon to the point of crunch; the positive-energy particle flies off to a distance. The energy it carries with it comes in the last analysis from the black hole itself. The massive object is no longer quite so massive because it has had to pay the debt of energy brought in by the negative-energy member of the newly created pair of particles.

Radiation of light or particles from any black hole of solar mass or more proceeds at an absolutely negligible rate—the bigger the mass the cooler the surface and the slower the rate of radiation. The calculated Bekenstein-Hawking temperature of a black hole of three solar masses is only 2×10^{-8} degrees above the absolute zero of temperature. The total

Stephen W. Hawking

Born January 8, 1942, Oxford, England. Hawking today holds the Lucasian Professorship at Cambridge University, to which Isaac Newton had been appointed in 1669. Hawking pointed out as early as 1972 that "black holes formed by collisions of smaller black holes can undergo further collisions, releasing more energy, whereas matter that has been fully processed by nuclear reactions cannot yield any more energy by the same means." He is an inspiring leader in the search for new insights at the frontiers between quantum theory, astrophysics, and geometrodynamics.

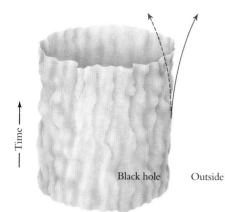

The Hawking evaporation process capitalizes on the fact that space is nowhere free of so-called fluctuations, evidence that everywhere latent particles await only opportunity—and energy—to be born. Associated with such fluctuations at the surface of a black hole, a might-have-been pair of particles or photons can be caught by gravity and transformed into a real-life particle or photon (solid arrow) that evaporates out into the surroundings and an antiparticle or counter-photon (dashed arrow) that "goes down" the black hole.

thermal radiation calculated to come from its 987 square kilometers of surface is only 1.6×10^{-29} watt, less than one-billionth of one-billionth of one-billionth of a watt! A black hole of any substantial mass is thus deader than any planet, deader even than any dead moon—when it stands in isolation.

Lighthouses of the Universe

In contrast to dead solitary black holes, the most powerful source of energy we know or conceive or see in all the universe is a black hole of many millions of solar masses, gulping down enormous amounts of matter swirling around it. Maarten Schmidt, working at the Mount Palomar Observatory in 1956, was the first to uncover evidence for these quasistellar objects, or quasars, starlike sources of light located not billions of kilometers, but billions of *lightyears* away. Despite being far smaller than any galaxy, the typical quasar manages to put out more than 100 times as much energy as our own Milky Way, with its hundred billion stars. Quasars, unsurpassed in brilliance and remoteness, we count today as the lighthouses of the heavens.

Observation and theory have come together to explain in broad outline how a quasar operates. A black hole of some hundreds of millions of solar masses, itself built by accretion, accretes more mass from its surroundings. The incoming gas, and stars-converted-to-gas, does not fall in directly, anymore than the water rushes directly down the bathtub drain when the plug is pulled. Which way the gas swirls is a matter of chance or past history or both, but that it does swirl is a straightforward consequence of the law of conservation of angular momentum. This gas, as it goes round and round, slowly makes its way inward to regions of ever stronger gravity. Thus compressed, and by this compression heated, the gas breaks up into electrons—that is, negative ions—and positive ions, linked by magnetic fields of force into a gigantic accretion disk. Matter little by little makes its way to the inner boundary of this accretion disk and then, in a great swoop, falls into the black hole, on its way crossing the horizon, the surface of no return. During that last swoop, hold on the particle is relinquished. Therefore, the chance is lost to extract as energy the full 100 percent of the mass of each infalling bit of matter. However, magnetic fields do hold onto the ions effectively enough for long enough to extract, as energy, several percent of the rest mass. In contrast, neither nuclear fission nor nuclear fusion is able to obtain a conversion efficiency of more than a fraction of a percent. Of all methods to convert bulk matter into energy, no one has ever seen evidence for a more effective process than accretion into a black hole,

Chapter Twelve

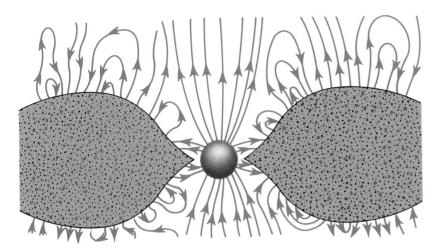

Magnetic lines of force (green) link the chaotic swirl of positive and negative ions in the rotating accretion disk (orange). The black hole by its overpowering influence lines up the lines of force near its center, where they become a guide for jets of particles ejected from the poles of the black hole along its axis of rotation.

and no one has even been able to come up with a more feasible scheme for one.

The magnetic field that is powered by the infalling matter indirectly generates an electric field and thus a voltage at the poles of the rotating black hole of the order of magnitude of 10^{20} volts. These voltages and fields generate jets of matter that drive straight out through surrounding interstellar gas for hundreds of thousands of light years and more. During a hundred thousand or million years, the accreting stars and gas can hardly be expected to have kept always to the same plane of accretion. Therefore, one might at first expect photographs of the luminous jet trail to be as wobbly as the stream of water from a hose held in an unsteady hand. But the trail isn't wobbly. It's straight. No other natural explanation offers itself but this: the black hole dominates. It is a gyroscope that holds the jet straight. Fluctuations from year to year in the swirl of the accreting matter do not upset it because it is so massive and has gathered so much angular momentum from a so much greater quantity of matter that it has accreted over so many millions of years.

Of all features of black hole physics in action, none is more spectacular than a quasar. And no lighthouse of the skies gives more dramatic evidence of the scale of the universe.

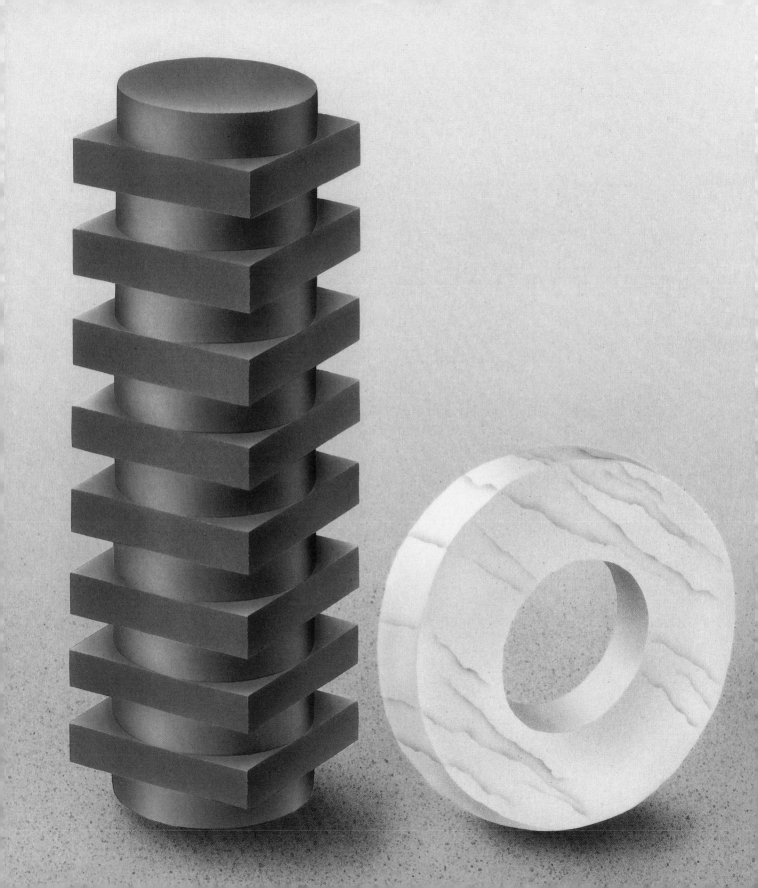

13 *A Farewell Look at Gravity*

It looks strange,
And it looks strange,
And it looks very strange.
And then suddenly
It doesn't look strange at all,
And you can't understand
What made it look strange
In the first place.

— GERTRUDE STEIN, *on modern art*

The mysteries concealed by the man-invented terms, time and space, are symbolized here by two ancient Chinese forms.

Only by understanding gravity as grip of spacetime on mass, and mass on spacetime, can we comprehend even the first thing about the machinery of the world—the inertia of a particle, the motion of the planets, the constitution of a star. Without a grasp on gravity, we would not perceive the link between the fall of the nearest pebble, the orbit of Europa, the beelike buzzing of a star cluster, the two hundred million years of each tour of the Earth about the Milky Way, and the fantastic power output of the distant quasars. We would find ourselves perplexed by the doubling of the telescopic image of a distant quasar, by the expansion of the universe and the gradual slowing of this expansion, and by the primordial process that manufactured the particles and the elements. Thanks to gravity and astrophysics, we now have some understanding of every one of these processes. To gain so much from gravity, however, does not give us everything. Existence, the preposterous miracle of existence, still remains the mystery of all mysteries. Even the word *universe*, bandied about in many a book, conceals a mountain of ignorance. We don't know whether what we call "the universe" is open or closed or even whether that distinction—with growth in our understanding—may not itself fade away into thin air, into undefinability, into nonsense.

To trace out what we know today about all the applications and ramifications of gravity, the evidence for them that lies before us unexplained, the questions that dog us, and the tools that we have to explore them, would take volumes. In concluding this little book on the "guts of gravity" let us take a fleeting look at frontier topics as a happy way to say to gravity, "Farewell for today. Until tomorrow!"

The Expanding Universe

No topic will serve more simply and more quickly as point of departure for discussing many of the processes governed by gravity than the expansion of the universe. This expansion we see in the ever widening distance between clusters of galaxies. To picture the expanding universe, we partially inflate a balloon, paste pennies on its surface to represent galaxies, and blow up the balloon to a greater size. Each penny sees the pennies around it receding from it. Moreover, the pennies twice as far away recede twice as fast.

A high-school student once asked me a question that typifies the response of many persons on first hearing about the idea of an expanding universe. "If the universe is expanding, as you say, then is everything in the universe, including ourselves, also getting bigger? If not, what takes the place of the big spaces in the universe that result from its expansion?" This student's question reflects a common misunderstanding: that is, if

Over a thousand gravimeters like this, in daily use throughout the world, make important contributions to the location of underground oil deposits and other minerals. Adjustments of the dial (white) bring a spring-supported mass, inside the box, into a balance checked by the microscope (black) whose eyepiece rises just beyond the dial. This LaCoste and Romberg gravimeter is capable of distinguishing accelerations of gravity that differ as little, for example, as 980.12345 and 980.12347 centimeters per second per second. Minor differences in surface gravity between point and point over stretches of landscape with horizontal separations of the order of 100 to 1000 meters, when corrected for change of altitude and for Moon- and Sun-driven tide-producing accelerations, reveal density differences in the underlying geologic structure.

the distance between one cluster of galaxies and another expands, then the distance between Sun and Earth expands, the length of a meter stick expands, the diameter of every atom also expands. But if all that were true, it would make no sense to speak of any expansion at all. Expansion relative to what?

In the expanding universe, *only* the distance between one cluster of galaxies and another expands. Atoms don't expand. Meter sticks don't expand. We don't expand. The balloon with its firmament of pennies provides a happy mental picture of the expanding universe. As the balloon expands, the distance between penny and penny expands, but not one penny itself expands.

One deficiency of the balloon model is hard to remedy. An ant standing on its penny sees itself surrounded by a mere circle of other pennies, not by a whole sphere of heavenly bodies as we do. Only one parameter of direction, only one angle does the ant need to specify the direction of a particular neighbor, and one more parameter, a distance, to complete the specification of its location. The ant's world is two-dimensional. Our own world is three-dimensional. In it we require two parameters of direction, two angles, to tell our telescope which way to point to see a particular galaxy; and a third parameter, a distance, to tell how far away it is.

I have never talked with a balloon-bound ant. I do not know whether it conceives, or even can conceive, of its unbounded but closed 2-sphere rubber world to be embedded in a flat 3-space. But we, surely, need only do a little hop, skip, and jump of the imagination to think of the 3-sphere universe of galaxies to be embedded in a flat 4-space. (This flat four-dimensional *space* has nothing whatsoever directly to do with our real physical world of *spacetime*, even though that also is four-dimensional.) Almost all of this flat 4-space is totally out of our reach, quite untouchable, pure talk: the center of the embedded 3-sphere is utterly inaccessible, and so is the whole of its interior and all the endlessness outside it. Only the 3-sphere itself is real. It alone stands for the space in which we live and move, according to the Einstein-Friedmann model of a closed but unbounded universe.

I confess I have never been able, myself, to picture directly this mythical, flat, infinite 4-space and the 3-sphere universe embedded in it. In my thought I always drop a dimension. I come back to seeing myself in the world of the ant, living on the surface of a balloon, able to travel in any direction in a space that has limited extent but no limits. Einstein taught us to think of the universe in these terms, closed but unbounded. The only natural boundary condition, he argued, is *no* boundary!

If the size of the Einstein-Friedmann model universe were never to change, then in it we could proceed in any direction, go straight on with never a swerve of direction for an immense distance, and find ourselves back at our starting point. But it does change! The universe expands.

It took Einstein fifteen years to recognize and accept the expansion. His greatest hero and influence was Spinoza, who had been excommunicated from the Amsterdam synagogue in the early sixteen hundreds for denying the biblical account of a moment of creation. Where in all that nothingness before creation would the clock sit that would give the signal to start? So his reasoning ran. Based on this logic, Einstein concluded that the universe must run on from everlasting to everlasting. His great geometric theory of gravity, however, provided no natural model of the universe that would meet this requirement. Therefore, he introduced into the basic principle of the theory—the principle that momenergy governs moment of curvature—the least unnatural change he could, a so-called cosmological term whose whole point and purpose was to provide a static universe. Then came 1929 and the discovery by Edwin Hubble at the Mt. Wilson Observatory that the universe is not static; it is expanding. Galaxies twice as far away as galaxies closer at hand move away from us at twice the speed. Such an expansion had already been predicted by Alexander Friedmann in 1922 on the foundation of Einstein's original simple theory. Each galaxy thinks it is the center of expansion, just as each penny on the enlarging balloon, with equal justice— and equal self-centeredness—thinks of itself as the center of expansion. After Hubble's discovery, Einstein rated the idea of a cosmological constant as "the biggest blunder of my life." Nevertheless, all credit goes to the original 1915 theory and to the man who gave it to us. For it to have predicted correctly and against all expectation a phenomenon so fantastic as the expansion of the universe—when was mankind ever granted a more wonderful token that we will some day understand existence itself!

Many times in the half-century and more since Hubble's discovery, others besides the founding father himself lost faith in the simple straightforward predictions of Einstein's standard 1915 geometrody-

Relation between red shift and distance for extragalactic nebulae

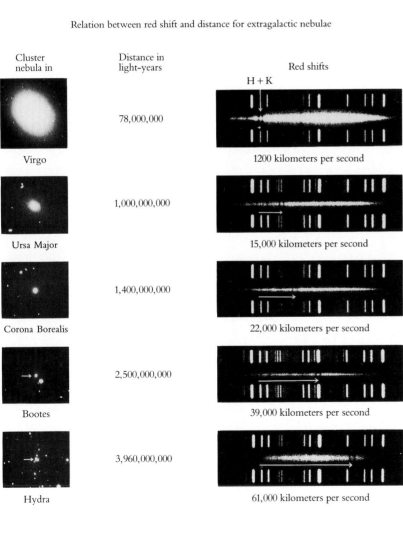

Cluster nebula in	Distance in light-years	Red shifts
Virgo	78,000,000	1200 kilometers per second
Ursa Major	1,000,000,000	15,000 kilometers per second
Corona Borealis	1,400,000,000	22,000 kilometers per second
Bootes	2,500,000,000	39,000 kilometers per second
Hydra	3,960,000,000	61,000 kilometers per second

Evidence from known distances on the expansion of the universe and the relation between red shift and distance is used here to determine the distances of selected extra-galactic nebulae. On the left appears the object studied in the telescope. A spectroscope operates at the point where the light from this object falls on the focal plane of the telescope. The spectroscope compares the spectrum of the object (central horizontal band in the pictures at the right) with the standard so-called H and K lines of the calcium ion. The fractional shift toward the red reveals immediately the velocity of recession (arrow) of the source relative to the velocity of light. From this recession velocity we can figure the distance by way of the tentative figure for the Hubble expansion rate of 50 kilometers per second per million parsecs, that is, 17 kilometers per second per million light years.

GRAVITY'S NEXT PRIZE: GRAVITOMAGNETISM

As electric charge, going round and round in a circle, produces magnetism, so mass, going round and round in a circle, must produce a new kind of force, gravitomagnetism. Oersted discovered the creation of magnetism from moving electric charges in 1820, but gravitomagnetism has not yet been discovered, although it is the target of two great research enterprises. Of all ways to detect this new effect, the most promising relies on the Einstein-Lense-Thirring–predicted turning of the axis of a gyroscope under the influence of gravitomagnetism. The quickest way to an order of magnitude estimate of this Einstein-Lense-Thirring "frame dragging" is a line of reasoning invented by Mach which was very influential in guiding Einstein to his geometrodynamics. In essence, Mach's principle states that inertia here arises from mass there. Einstein's theory transmuted Mach's bit of philosophy into a bit of physics, susceptible to calculation, prediction, test. It is not necessary to enter into the mathematics of the theory to state its simple consequence for simple cosmologies. Each mass has an "inertia-contributing" power, a voting power, equal to its mass, there, divided by the distance from *there* to *here*.

Let's look at this idea in action! Mount a gyroscope on frictionless gimbals or, better, float it weightless in space to escape the gravity that on Earth grinds surface against surface. Picture our ideal gyro as sitting on a platform at the North Pole. Pointing initially to the flagpole at a corner of the support platform, will the gyro continue to point that way as a 24-hour candle gradually burns away? No. Its axis of spin points and continues to point to one star as the day wears on. The Earth turns beneath the heedless gyro, one rotation and many another as the days go by. The inertia–determining power of the mass spread throughout space is seen in its action on the gyro.

Wait! As conscientious workers against election fraud, let's inspect the ballot that each mass cast. Where is the one signed "Earth"? Surely it was entitled to vote on what would be the free-float frame at the location of the gyro. The vote tabulator sheepishly confesses, "I know that the Earth voted for having the frame of reference turn with it. However, its vote had so little inertia-determining power compared to all the other masses in the universe that I left it out."

"Tut tut," I admonish him, "you did wrong. Let's see how much damage you have done." Yes, the voting power of the Earth at the location of the gyro is small, but it's not zero. It is given by the expression (mass of earth/radius of earth) = (0.44 centimeters/6.73 × 10^8 centimeters) = $0.65 × 10^{-9}$, roughly only one-billionth as much influence as all the rest of the universe together."

This 4-centimeter-diameter quartz-sphere gyroscope, heart of the Stanford gravitomagnetism experiment, spins at 10,000 revolutions per minute, centered to a few millionths of a centimeter within an evacuated quartz housing. It is designed to float in orbit 600 kilometers above Earth in the satellite Gravity Probe B and to measure a change in the direction of axis of spin to a precision of a thousand times better than the angular resolution of a typical, very good star-tracking telescope.

A model of the Laser Geodynamic Satellite (LAGEOS), launched in 1976 and since then tracked by 30 laser-equipped stations scattered over Earth's surface. Its position is regularly read with an uncertainty less than a few centimeters—a precision better than that of any other object in the sky. A second such satellite going around the Earth with an opposite inclination has an orbital axis that will undergo a like "large" precession caused by Earth's nonsphericity, but an opposite small gravitomagnetic precession. Thus the difference between the two turning rates will measure a force of nature still new to man.

"It wasn't only because its voice was so weak that I threw out the Earth's vote," the tabulator adds. "The free-float frame of reference that the Earth wanted the gyro axis to adhere to differs so little from the frame demanded by the faraway stars. The earth wants the gyroscope axis to creep slowly around once in a 24-hour day rather than keep pointing at one star. Who cares about that peanut difference?"

"Again you are wrong," I tell him. "In the course of a whole year, the axis of the gyro, if it followed totally and exclusively the urging of the Earth, would turn through 473×10^9 milliseconds of arc relative to the distant stars. That's the turning effect you considered so miniscule that you threw away the Earth's ballot!"

The turning rate of the gyro, relative to the distant stars, can be calculated by multiplying the voting power of Earth (0.654×10^{-9}) by the rate of turn desired by Earth (473×10^{-9} milliseconds of arc per year), to give a result of 309 milliseconds of arc of precession per year for a gyroscope floating just above the surface of the Earth. What about a gyro floating two Earth radii from the center of the planet? Because a sphere of such dimensions has eight times the volume of the Earth, the magnitude of the sought-for force is diluted down to about one-eighth its value at the surface of the Earth. Rough and ready though the reasoning is that leads to this estimate, it gives the right order of magnitude.

Nobody has figured out how to operate on the Earth's surface a gyroscope sufficiently close to friction-free that it can detect the predicted effect. In both projects aimed at detecting gravitomagnetism, the detecting gyroscope will be in free-float orbit above the Earth. In the Stanford project, the gyroscope is a quartz sphere (facing page), smaller than the hand, that is carried along in orbit with an air-cushion suspension and sensitive read-out devices. In the LAGEOS project, the gyroscope is huge, roughly four Earth radii in diameter. It consists of a 60 centimeter sphere covered with 426 mirrors, going round and round in orbit every 3.7518 hours. An axis perpendicular to that orbit is identified as the gyroscope axis. For it, the predicted gravitomagnetic precession is 31 milliseconds of arc per year, quite enough to be detected if there were not an enormously larger precession at work that drowns it out. This devastating competition, amounting to about 126 degrees per year, arises from Earth's nonsphericity. Fortunately the possibility exists to launch another LAGEOS satellite going around at another inclination, so chosen that the anomalies in the Earth's shape give the same orbit turning as before but the Lense-Thirring effect is reversed.

Gravity Probe B, *the space vehicle built to carry into weightlessness the Stanford gyroscope of page 232. The gyroscope is designed to roll with a 10 minute period around the line of sight to the star Rigel. The direction of the spin axis of the gyroscope will be compared to that line of sight throughout the year to detect and measure the gravitomagnetic "frame dragging" caused by the whirl of Earth's mass.*

namics. Many a radical cosmological concept has been put forward to explain observational findings that later proved to be mistaken or misinterpreted or both. Among such crises, none has endured longer or led to more numerous sound astrophysical observations and careful theoretical analyses than what was termed already in 1958 "the mystery of the 'missing' mass."

The Mystery of the Missing Mass

The mystery of the missing mass takes four forms: (1) Where is the mass needed to hold our Sun and other stars in their multimillion-year orbits about the center of the Milky Way? We don't see it! (2) Where is the mass that holds the other galaxies together as well? (3) Where do the pulls or pushes come from that—on the cosmological scale—assemble galaxies into great filaments separated by great voids? And, (4) where is the mass that is needed to curve up the universe into closure? We don't see it, either. The answers to these four questions are still being worked out in that marvelous world-wide collaboration that is science, in day-by-day

Chapter Thirteen

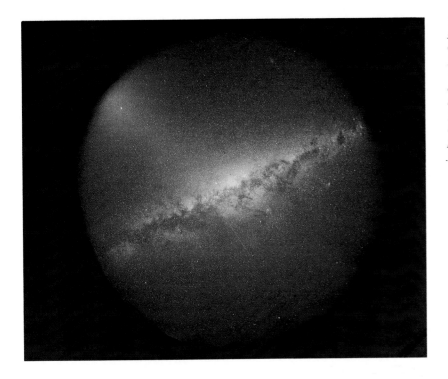

The Milky Way, the nebulous band that stretches around the sky, recognized by Galileo in March 1610 to be stars and by Immanuel Kant in 1755 to be held in orbit "through their mutual attraction." Kant went on to say, "Ruin is prevented by the action of the centrifugal forces . . . the same cause [centrifugal force] . . . has also so directed their orbits that they are all related to one plane . . . [turning] one degree [or less] in four thousand . . . years."

consultations between colleagues, week-by-week journal publications, and season-by-season conferences. Already at this writing, however, it looks more and more as if the answer to all four questions may be the same: almost all of the missing mass sits out there right now as plain ordinary matter, not yet aggregated into assemblies large enough to have become luminous.

Let's look at the evidence on individual galaxies and assemblies of millions of galaxies before we turn back to the issue on which they all bear—the mass and fate of the universe.

Galileo once described our galaxy, the Milky Way, as "nothing else than a mass of luminous stars . . . set thick together in a wonderful way." The galaxy has the flattened form of a fried egg. Our solar system lies about at what would be the division line between yolk and white, some 30,000 lightyears from the center. At the center of the yolk is a group of stars, visible by telescope, in rapid circulation about a central concentration of mass, believed to be a black hole of about 3.5 million times the mass of the Sun. It takes the far larger mass of the galaxy itself to hold the stars together—by gravitational attraction—in slow revolution about this center. First fully to appreciate this point and to present it in a book was Immanuel Kant (1724–1804). However, after his publisher went bankrupt, the chief creditor locked up all copies of Kant's book in a warehouse for years and years; thus his findings did not greatly influence the course of science. Otherwise his fame as an astrophysicist might have been comparable to his fame as the great philosopher of Königsberg.

Kant did not have the tools available to work out the details. Today we do. The mass of the stars observed to lie inside the orbit of the Sun adds up—within 10, 20, or 30 percent—to what is needed to hold the Sun in its orbit (with its observed orbital radius and observed speed, relative to the speed of light), as figured from the Kepler 1-2-3 law in the form:

$$\text{mass in meters} = \text{orbital radius in meters} \times (\text{speed})^2$$

For stars further from the center of the galaxy than the Sun, however, the deficiency of mass observed compared to mass required to hold those stars in their observed orbits grows greater, amounting to a factor of two or more.

It is easier in some ways to apply the Kepler 1-2-3 law to other galaxies than to our own. Stars on one side of an elliptic galaxy, seen edgewise, move toward us; on the other side, away. To plot out this

The central region of the "nearby" cluster of galaxies in the constellation Fornax. The typical galaxy contains on the order of 100 billion stars.

Chapter Thirteen

velocity of selected stars and its dependence on their distance from the center is to come up with what is generally termed a "velocity curve" such as the two illustrated here from the two galaxies NGC 2403 and NGC 3198. If the only source of gravitational attraction were the objects visible in the telescope, the pulling power of the galaxy, far out, could only hold a star in orbit at that distance if that star went around at something like half the speed it is observed to have. Conclusion? There is far more mass than what we see helping to hold that star in orbit!

One valuable clue to the mass of a distant galaxy or cluster of galaxies—called *gravitational lensing*—is beginning to be exploited. No one will feel baffled by this phenomenon who has looked at a distant street light through a pane of glass afflicted by a lenslike imperfection and seen a double or triple image. Likewise, a double or triple telescopic image is occasionally observed of an extremely distant, powerful, and concen-

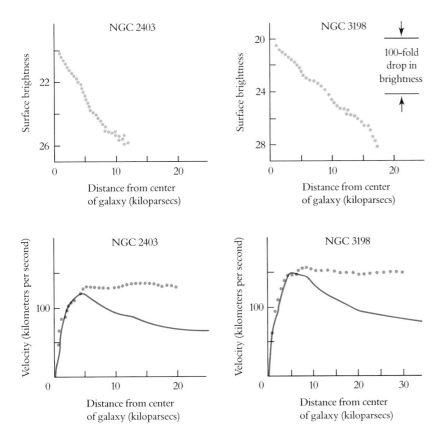

Top: *Surface brightness of two well-studied galaxies, each shown in its dependence on distance from the center of that galaxy. One kiloparsec is 3262 lightyears distance. On the scale at the left, magnitude 24, for example, is tenfold dimmer in surface brightness than magnitude 22.* Bottom: *Evidence that the luminosity in the outer reaches of these galaxies gives no true measure of the mass present there. In those distant parts of the galaxy, the actually observed velocities of orbiting stars (blue dots) are higher than the velocities that would be calculated (red) if the only mass available to do the holding in orbit came from the stars whose light is seen (orange dots in top graphs). Evidently there is much mass that optical telescopes don't see.*

trated source of light known as a quasistellar source, or *quasar*. This happens when the quasar lies in line with a closer, but still very distant, galaxy or cluster of galaxies, which provides gravitational pull. This pull takes rays of light that are diverging from the original source and causes them to converge—to arrive at the same telescope from different directions. The photographic plate registers the quasar as double, triple, or multiple. This gravitational lensing action provides a tool (not yet honed to precision!) to get at the mass of the intervening galaxy or cluster.

The mystery of the missing mass grows greater (and therefore closer to solution!) when we turn from the motions within one galaxy to the message of millions of galaxies. Take a telescope that has a 6 degree opening, swing it through the sky through an arc of 135 degrees, and chart the speed of recession (or, equivalently, the distance from us to it, according to Hubble's law connecting distance with speed) for every galaxy in that pie-shaped slice of space that (1) has a visual magnitude of 15.5 or brighter and (2) is receding from us at a speed of 15,000 kilometers a second or less. That's the task V. Lapparent, Margaret Geller, and J. P. Huchra set themselves. One diagram on the facing page shows their spectacular 1986 finding: great voids and great filaments in the distribu-

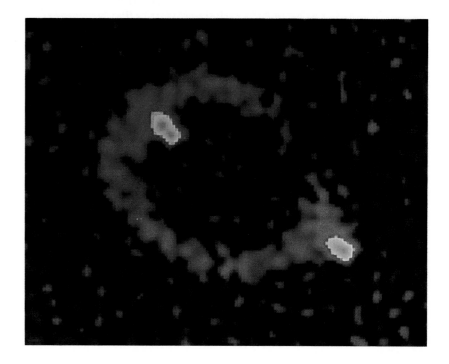

This ringlike pattern of radio-wave emission known as MG 1131 + 0456 was observed in 1987 by the Very Large Array (VLA) radio telescope, 50 miles west of Socorro, New Mexico. No reasonable explanation for this structure has ever been found except gravitational lensing of radiation from a very distant pointlike quasar source bent by the gravity of a foreground massive object, too faint to show here.

tion of the galaxies. The first reaction of many was astonishment: what strange pushes and pulls there must be at work in the universe to produce these inexplicable zones of clustering and zones of avoidance. Others, however, suspected there was nothing more needed to explain the broad features of this distribution of galaxies than plain ordinary gravity acting between plain ordinary masses. The most decisive evidence for this conservative interpretation reached me a few hours before this final revision of Chapter 13 from Changbom Park, a graduate student at Princeton. Park made a printout of his computer findings especially for this book. It shows an amazing similarity in character with the observations. Again clustering, again voids. Yet into those calculations went nothing but gravity. To look at what went into Park's analysis is to appreciate something of what the expanding universe is and means and does.

The calculations traced out the motion of two million mass points, each changing its location and speed from instant to instant in response to the Newtonian gravity of all the others. The number of points thus considered, unsurpassed in any such analysis ever made to date, is nevertheless sufficiently small compared to the scale of the universe as a whole that it is legitimate to neglect in the analysis all large-scale curvature of space. The points are followed from the very earliest time when, it is generally believed, todays' whole visible universe occupied a space far smaller in dimensions than an elementary particle. A natural line of reasoning, based on modern quantum theory, indicates that the initial distribution of mass, though fantastically uniform, must nevertheless have presented percentagewise exceedingly small but predictable fluctuations from this ideal uniformity. Starting with this initial condition, the calculations reveal that two million "tracer points," each of galactic mass, gradually move, under gravity alone, into remarkable associations and avoidances. This effect is exaggerated, both in nature and in the calcula-

Left: Observed distribution of galaxies of magnitude 15.5 and brighter. Included are those galaxies that lie in a selected pie-shaped band of direction whose speed of recession from us extends to fifteen thousand kilometers per second. Right: Changbom Park's calculations based on the so-called standard bias principle and a density of matter equal to that required to curve up the universe into closure.

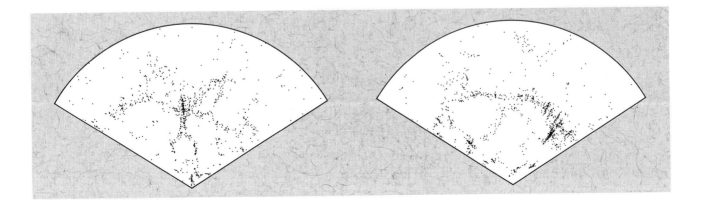

tions, by a built-in bias: those places where the tracer mass points come closer together symbolize locations where visible galaxies come into being; places where the tracer masses are further apart are locations where matter does not assemble itself into stars, and therefore these places will not be visible to any telescope. In other words, far more points went into Park's calculations than show on the diagram. In brief, gravity, plus the physical requirements for the turning of matter into visible stars, bias telescopes against seeing any matter at all in great regions of space where matter has a density a little lower than elsewhere. Only when we recognize and come to terms with this bias can we hope to understand even the first thing about the gravitational dynamics of the biggest system of all, the universe itself.

The Universe: From Big Bang to What?

Somewhere between ten and twenty billion years ago, today's galactic recession velocities indicate, the universe began with a big bang. Small in the early days, the universe had high space curvature and needed a high density of energy to bend it up into that tight ball. Large today, it has a far smaller space curvature and needs much less density of mass to bend it up into closure—yet a density five to ten times greater than any visual observations directly evidence. The bias principle tells us, however, that we no longer have to worry about the mystery of the missing mass. Optical telescopes see only luminosity, and luminosity gives a quite unfairly low count of mass in regions of a little lower than average mass density.

As relics of the big bang we see around us not only today's rapidly receding galaxies but also today's abundance of the chemical elements—some among them still radioactive, the "still warm ashes of creation" (V. F. Weisskopf)—and today's greatly cooled but still all pervasive "primordial cosmic fireball radiation." We now believe that in the first instants of its life, the entire universe filled an infinitesimally small space of enormous density and temperature where matter and energy fused in a homogenous soup. Immediately the universe began expanding. After about 10^{-6} seconds it had cooled enough that subatomic particles condensed from the matter-energy soup. In the first three minutes after the big bang, neutrons and protons combined to make heavier elements. Eons later stars and galaxies formed. Never since has the universe paused in its continual spread outward.

Will the universe continue expanding forever? Or will its expansion slow, halt, and turn to contraction and crunch, a crunch similar in character but on a far larger scale than what happens in the formation of a black hole? For insight on this question, consider a Moon floating in the emptiness of space. Plant explosives within it and explode it. If we plant too much explosive, supply too much energy, start off the expansion with too much speed, the Moon is gone for keeps. If we implant less than the critical explosive charge, then contraction and final crunch will inevitably follow the initial bang. A great deal less charge, then a relatively short time between bang and crunch. A charge only moderately less than the critical amount, then a longer time before the outcome declares itself. But to tailor the explosive charge so delicately that the eventual outcome—expansion forever or contraction and final crunch—declares itself only after many billions of years seems to require such fine-tuning as to be incredible.

Right reasoning about the Moon has occasionally led to wrong reasoning about the universe. Some argue that the idea of a closed universe still expanding, though at a slowed rate, many billions of years after the initial big bang is preposterous. No one starting the whole show at time zero—it is sometimes said—could possibly adjust the initial propulsive energy with sufficient finesse to make things come out just right so far in the future. That mistaken argument is sometimes put forward to justify belief in, search for, and proclamation of one or another new idea about how things started, some new principle that will automatically guarantee the desired delicacy of adjustment of the energy. But Einstein's geometric theory of gravity tells us that no adjustment of energy needs to be made or even can be made.

One mass? One radius at maximum expansion, one time from bang to crunch. Twice that mass? Then twice the radius at maximum expansion, twice the time from bang to crunch. No tailoring of the "explosive charge." No fine-tuning of the initial expansion rate. No worry for fear that the closed model universe will go on expanding forever. The total mass of those clusters of galaxies, those "rocks," that make up the closed universe totally fixes the time scale of the expansion-plus-contraction, the size of the universe at the phase of maximum expansion, and the size and rate of expansion at any specified time after the big bang. The table on the following page displays some of the features of a 3-sphere closed universe compatible with astronomical observations.

The numbers in the table are at most illustrative. There are several reasons for caution about them. First, even today astronomers of the greatest skill and integrity have difficulty with the determination of galactic distances. Different groups may arrive at a distance for a particular faraway galaxy that differ from each other by a factor of two. The resulting uncertainty in the distance scale carries with it corresponding uncertainties in the size of the Friedmann model with which it should be compared and in the total mass. Second, close to the times of the big bang

A closed model universe compatible with observation

Radius at phase of maximum expansion:	18.9×10^9 lightyears or 1.79×10^{28} cm
Time from start to maximum size:	29.8×10^9 years or 2.82×10^{28} cm
Radius today:	13.2×10^9 lightyears
Time from start to today's size:	10.0×10^9 years
Time it would have taken from start to today's size if the entire expansion had occurred at today's slowed rate of expansion:	20.0×10^9 years
Present expansion rate:	An extra increment of recession velocity of 49.0 kilometers per second for every extra megaparsec (equivalent to 3.26 million lightyears) of remoteness of the galactic cluster
Fraction of the way around the 3-sphere universe from which we can in principle presently receive light:	$\dfrac{113.2 \text{ degrees}}{180 \text{ degrees}} = 62.9\%$
Fraction of the matter in the 3-sphere that has been able to communicate with us so far:	74.4%
Number of new galaxies which, on average, come into view every three days:	One!
Average mass density today:	14.8×10^{-30} grams per cubic centimeter
Average density at phase of maximum expansion:	5.0×10^{-30} grams per cubic centimeter
Rate of increase of volume today:	1.82×10^{68} cubic centimeters per second
Amount of mass:	5.68×10^{56} grams; in geometric units, 4.21×10^{28} centimeters
Equivalent number of suns like ours:	2.86×10^{23}
Equivalent number of galaxies like ours:	1.6×10^{12}
Equivalent number of baryons (neutrons and protons):	3.39×10^{80}
Total time, big bang to big crunch:	59.52×10^9 years

and the big crunch, radiation becomes so hot and energetic that it contributes more than matter to the density of mass in the universe, with resulting minor corrections to some of the numbers listed. Finally, we have no absolute guarantee that the universe is indeed closed, or whether a deeper, quantum-based understanding of existence will still leave meaningful the concept of "space closure."

Einstein's battle-tested and still-standard geometric theory of gravitation leaves no room for ifs, ands, or buts about the dynamics of a 3-sphere model universe. It is a one-cycle only dynamics. It totally rules out any cycle after cycle after cycle history, world without end. No place whatsoever does it provide for any "before" before the big bang, any "after" after the big crunch. Never has there been a human without a navel. No other way offers itself for his creation. It is conceivable that the universe likewise could never have come into existence were it not endowed in its history-to-be with a double belly button: big bang and big crunch. What goes on at those two gates of time is still, however, beyond our ken.

We have come to the end of our journey. We have seen gravity turn to float, space and time meld into spacetime, and spacetime transformed from stage to actor. The grip by which spacetime acts on mass and mass on spacetime turns out not to be some elaborate gadgetry built of gears and pinions, but a mathematical principle of incredible simplicity—the principle that the boundary of a boundary is zero. Of all the indications that existence at bottom has a simplicity beyond anything we imagine today, there is none more inspiring than the unsurpassed simplicity of gravity, as we now see it. Someday, surely, we will see the principle underlying existence as so simple, so beautiful, so obvious that we will all say to each other, "Oh, how could we all have been so blind, so long!"

I express special gratitude to the Princeton and Austin communities for the magic air of theoretical physics that floats about both places and for the privilege both have given me to learn by teaching challenging subjects to challenging students. I thank Edwin F. Taylor, co-author of our undergraduate text, *Spacetime Physics,* and Charles W. Misner and Kip S. Thorne, co-authors with me of the graduate text *Gravitation,* both of them Freeman books, for all I learned working with them. For the original invitation to undertake the present book, I express warmest appreciation to Edmund F. Immergut and his advisors, Gerard Piel and Philip Morrison. Peter Renz and Shaughan Lavine gave me most helpful editorial assistance on the early drafts of the book, Aidan Kelly gave marvelous guidance on cuts and clarity, and Susan Moran and Rita Gold were models of wisdom and taste in winding up the final stages of the editorial enterprise. I thank Mike Suh and his art team, including Tom Cardamone and Tomo Narashima for their skill in converting sketches into drawings and ideas into pictures, Travis Amos for locating hard to find and unusual photographs, Bill Page and Mara Kasler for processing the illustrations, Julia De Rosa for juggling a breakneck schedule and coordinating the production, and Louise Ketz for preparing the index. In thanking those I name, I thank also the whole Freeman *Scientific American Library* team and its skipper, Linda Chaput, for the spirit of excellence and leadership to which they adhere.

So many colleagues, correspondents, and friends have helped that I can do no more than wave to most an airborn "thank you." Special appreciation goes to duPont friends—too many, alas, no longer living—from Wilmington, Richland, and Hanford days for their support of early postwar research at Princeton; and to Jane and Roland Blumberg of Seguin, Texas, and to E. Bruce Street of Graham, Texas, for their assistance to the University of Texas Center for Theoretical Physics, which has one fruition in the present pages. Thanks go to John Stachel, Boston University, for Einstein items; to Roy L. Bishop, Acadia University, Wolfville, Nova Scotia, for the chance to see tides that are real tides and for a wonderful photograph of them; to Moscow colleagues Vladimir Braginsky, Leonid Grishchuk, I. M. Khalatnikov, L. I. Sedov, and—no longer living—E. M. Lifshitz, Andrei Sakharov, and Yakov B. Zel'dovich for rich insights on gravity.

I am indebted to Princeton colleagues for advice and assistance, among them John Bahcall, Hafez Chehab, Marshall Clagett, Peter Cziffra, Ruth A. Daly, Robert H. Dicke, Freeman Dyson, Val Fitch, Wen Fong, J. Richard Gott, James A. Lechner, Robert P. Matthews, Elliott McGucken, W. Jason Morgan, Otto Neugebauer, Bohdan Paczynski, Mark Palmer, Changbom Park, P. James E. Peebles, Jean F. Preston, Martin Ryba, Benjamin Skrainka, Lyman Spitzer, Joseph H. Taylor,

Acknowledgments

Letitia W. Ufford, David T. Wilkinson, and (deceased) Valentine Bargmann, Albert Einstein, Kurt Gödel, John H. Gomany, and John von Neumann.

I likewise express appreciation for help and wisdom to Austin colleagues Roberto Bruno, Marshall Burns, Bryce DeWitt, Cecile DeWitt, Raynor Duncombe, Harry King, Richard Matzner, Eva Riley, Christopher Salin, Debra Saxton, Danielle Rappaport, Benjamin Schumacher, Michael Struensee, Karl Trappe, and Steven Weinberg.

Warm thanks also go to Jacob Bekenstein, Beersheva, Israel; Hans Bethe, Cornell University; Joan Warnow Blewett, American Institute of Physics, New York; Joan M. Centrella, Drexel University; Edoardo Amaldi (deceased), Ignazio Ciufolini, and Remo Ruffini, Rome; David Bernstein, Donna Cox, David Hobill, Raymond Idaszak, and Larry Smarr, University of Illinois at Urbana-Champaign; Paul Davies and S. Keith Runcorn, University of Newcastle-upon-Tyne; Marek Demianski and Andrej Trautman, Warsaw University; C. W. Francis Everitt and William Fairbank (deceased), Stanford University; James Faller, Joint Institute for Laboratory Astrophysics, Boulder, Colorado; Adalberto Giazotto and V. Montelatici, University of Pisa; Sigmund Hammer, University of Wisconsin at Madison; Steven Hawking, Cambridge University; the distinguished designer known to so many as Ishi, San Francisco; Carson Mark, Los Alamos; Bahram Mashoon, Columbia, Missouri; George T. McWhorther, Ekstrom Library, University of Louisville, Kentucky; Christopher A. Mejia, Cambridge, Massachusetts; Jan Oort, Leyden; Oliver Payne, National Geographic Society, Washington; Alfred C. Prime, Malvern, Pennsylvania; Asghar Qadir, Islamabad; George Rideout, Gloucester, Massachusetts; Abdus Salam and Dennis W. Sciama, Trieste; Thomas Siegfried, Dallas; Claudio Teitelboim, Centro de Estudios Cientificos de Santiago, Chile; Charles H. Townes, University of California at Berkeley; Brian Tucholke, Woods Hole Oceanographic Institute, Massachusetts; William Unruh, University of British Columbia, Vancouver; Ryoyu Utiyama, University of Osaka; Masami Wakano, Kyoto University; John McG. Wheeler, Ardmore, Pennsylvania; Clifford Will, Washington University, St. Louis, Missouri; Adrienne Wootters and William Wootters, Williams College, Williamstown, Massachusetts; and Paul W. Worden, Jr., Stanford University.

Inspired by the practical uses of gravity and gravimetry in prospecting, I would have devoted at least a chapter of this book to that subject had not the judgment of editors prevailed. Nevertheless, I thank here the wonderful Texas friends I made in exploring this topic; Jane and Roland Blumberg of Seguin; Sam Worden (deceased) of Houston and his wife, Helen; and John Fett and Lucian LaCoste of Austin. Knowing them, I find it no accident that most of the dozens upon dozens of gravimeters in daily use throughout the world are made either by LaCoste & Romberg Gravity Meters, Inc. of Austin or, according to the Worden design, by Texas Instruments.

For help in tracking down the story of Apollonius of Perga, his *Conics,* Alexandria, and the manuscript of the decisive proposition on the parabola, I am indebted to Marshall Clagett of the Princeton Institute for Advanced Study; Lochlain O'Raifeartaigh of the Dublin Institute for Advanced Studies; David James of Dublin's Chester Beatty Library and Gallery of Oriental Art; George Saliba, Columbia University; George McWhorter, Patterson Rare Book Collection, University of Louisville, Kentucky; and Dorothy Thompson of the Princeton Institute for Advanced Study. For putting together draft after draft of this unusual book, and for help and judgment on it, I owe special thanks to Ruth Bentley, Orono, Maine; Zelda A. Davis, Austin; Christa Mayer-Bohne, Walpole, Maine; and Emily Langford Bennett of Princeton.

I am grateful to the National Science Foundation for the assistance it has given to my research over the years, most recently via NSF grant PHY 245-6243 to Princeton University. I thank here most of all my fellow citizens, who, for love of truth, take from their own wants by taxes and gifts to nurture great centers of light and learning.

John Archibald Wheeler

Illustrations by Tomo Narashima
and Tom Cardamone Associates

FACING PAGE 1
National Portrait Gallery, London

PAGE 3
The Granger Collection

PAGE 5
Deutsches Museum

PAGE 7
The Bettman Archive

PAGE 10
Lotte Jacobi/Department of Media
Services, University of New
Hampshire

PAGE 18
left, NASA
right, NASA

PAGE 21
Guy Sauvage Vandystadt/Photo
Researchers

PAGE 24
United States Geological Survey

PAGE 25
left, portrait by Sustermans, Firenze
Gallery/Art Resource.
right, Ann Ronan Picture Library

PAGE 26
top, Archives of the Loránd Eötvös,
University of Budapest

PAGE 27
Courtesy of Robert P. Mathews,
Princeton University

PAGE 31
R. A. Perley, A. H. Bridle, A. G.
Willis/NRAO

PAGE 34
Kennan Ward

PAGE 36
Garrett Arabic Collection, Princeton
University

PAGE 52
National Institute for Standards and
Technology

PAGE 56
Michael Hanson and Christian
Taylor

PAGE 66
Kennan Ward

PAGE 90
Roy Bishop, Acadia University,
Wolfville, N.S., Canada

PAGE 92
Lawrence Berkeley Laboratory.
Color by David Parker.

PAGE 103
Fogg Art Museum, Harvard
University, Cambridge,
Massachusetts.
Bequest of Edmund C. Converse.

PAGE 104
U.S. Department of Energy

PAGE 119
Academie des Sciences, Paris

PAGE 122
NASA

PAGE 124
Bildarchiv Preussischer Kulturbesitz

PAGE 133
left, Susan Schwartzenberg/
Exploratorium

PAGE 143
Giraudon/Art Resource

PAGE 145
Chester Beatty Library and Gallery
of Oriental Art, Dublin

PAGE 164
Jet Propulsion Laboratory

PAGE 167
Adapted from R. V. Pound and
J. L. Snider, *Physical Review B 140,*
p. 788 (1965).

PAGE 178
Magnum

PAGE 186
The Granger Collection

*Sources of the
Illustrations*

PAGE 191
Westinghouse Historical Collection

PAGE 192
Cavendish Laboratory

PAGE 194
Adapted from figure 4.6 of C. W.
Misner, K. S. Thorne, and J. A.
Wheeler, *Gravitation,* W. H.
Freeman and Company, San
Francisco (1973).
Created by Claudio Teitelboim.

PAGE 203
National Center for Supercomputing
Applications

PAGE 204
bottom, California Institute of
Technology

PAGE 205
top, The Virgo Project, the
Dipartimento di Fisica
dell'Universita di Pisa, Adalberto
Giazotto, and V. Montelatici
bottom. Adapted from a graph by
Joseph Taylor

PAGE 208
Helmut Wimmer

PAGE 211
Harvard Smithsonian Astrophysical
Laboratory

PAGE 213
Michael Simpson

PAGE 214
Adapted from R. Ruffini and J. A.
Wheeler, *Relativistic Cosmology and
Space Platforms,* European Space
Research Organization, Paris, France
(1971).

PAGE 219
Adapted from J. A. Wheeler,
Accademia Nazionale Dei Lincei,
Quaderno N. 157 (1971).

PAGE 223
R. Cagnoni/Black Star

PAGE 225
Adapted from figure 36(c) of Kip S.
Thorne, Richard H. Price, and
Douglas A. Macdonald, *Black Holes:
The Membrane Paradigm,* Yale
University Press, New Haven, CT
(1986) which, in turn, is adapted
from D. A. Macdonald and W. -M.
Suen, *Physical Review D 32,* p. 848
(1985).

PAGE 228
LaCoste and Romberg Gravity
Meters, Inc.

PAGE 229
Adapted from figure 27.2 of C. W.
Misner, K. S. Thorne, and J. A.
Wheeler, *Gravitation,* W. H.
Freeman and Company, San
Francisco (1973).

PAGE 231
California Institute of Technology

PAGE 232
Michael Freeman

PAGE 233
NASA

PAGE 234
Barron Storey

PAGE 235
Dennis di Cicco

PAGE 236
ROE/AAT Board, 1984

PAGE 237
From Kornelis Begeman, "Hl
Rotation Curves of Spiral Galaxies,"
Ph.D. diss., University of
Groningen, 1987.

PAGE 238
J. N. Hewitt, E. L. Turner/NRAO

PAGE 239
left, from Margaret J. Geller and
John P. Huchra, p. 8 in V. C.
Rubin and G. V. Coyne, eds., *Large-
Scale Motions in the Universe: A
Vatican Study Week,* Princeton
University Press, Princeton, New
Jersey (1989).
right, Changbom Park

Index

Callisto, 180
Canopus, 46–52
Carnot, Sadi, 220
Cartan, Élie, **119** (*see also* Einstein-Cartan equation)
Carter, Brandon, 222
Cavendish, Henry, viii, 210
Center of attraction
    planetary motion, 174–177, 179–180
Chesterton, G. K., 190
Christodoulou, Demetrios, 216
Circumference, reduced (*see* Schwarzschild radial (*r*-) coordinate)
Clagett, Marshall, 142
Closed universe, **239,** 241–243
Collisions
    conservation of momentum and energy in, **94**–96, 101–106
Comets
    motion of, 170, 177
    orbits, 182, 183
Conics (Apollonius of Perga), 142, 143, **145**
Conservation
    of angular momentum, 168–172
    of angular momentum and planetary motion, 179, 183
    of energy in planetary motion, 170–171
    of momentum, 196
    of momentum and energy, 95–97, 102
    of momentum-energy (momenergy), 101–106, 113–114, 120
Contractile spacetime curvature, 82–88
Copenhagen Institute for Theoretical Physics, 187
Corona Borealis
    cluster nebula in, **231**
"Creation," measure of, 105
Crunch
    of particles and black holes, 214
    point of in black holes, 210
    of universe, 240–241
Curvature (*see also* Negative curvature; Positive curvature; Surface curvature; Zero curvature)
    double, 142–147
    of Earth, 5, **6, 24**
    Gauss on, 4–6
    in mass-free space, 128–132

radii of, 133–140, 142
radii of and parabola, 143–147
rafter, 131
Schwarzschild geometry, 124, 126, 137–142, 160
of space, 140, 241
space geometry, 150, 158
of spaces, 5–7
of spacetime, 9, 11–14, 22–24, 27–30, 64–65, 68–76, 81, 119, 120, 162, 195, 198
of spacetime and boomeranging, 56–59, 61–65
of spacetime and gravity waves, 201–203
of spacetime and mass, 114–115
of spacetime and orbit of Mercury, 182
of spacetime and time in orbit, 180
of spacetime and transport, 78–80
of spacetime geometry, 124–147, 150, 166
of spacetime inside and outside Earth, 82–88
in-the-surface, 136–137, 140–142
and swing, 132
tide-driving, 82, 84–88, 119
of worldlines, 69–70
Cygnus region, 211–212
Cygnus X-1, 211–213

Density
    of Earth, 211
    of Sun, 210
    of universe, 240, 242
    of water, 211
Detectors
    gravity waves, 204–205, 207
Dicke, Robert H., **27**
Dieudonné, Jean, **119**
Differential
    Schwarzschild space metric, 153–154
Dilution
    spacetime curvature, 82, 86–87
Directrix
    rafter, 131
    Schwarzschild geometry, 127, 131, 137–140
Distance
    and curvature, 80–81, 137–140
    differential of, 153–154
    between events, 33

between galaxies, 241–243
measuring, 36
metric formula measuring, 150–159
and moments of rotation, 130–131
and red shift, **231**
Schwarzschild radial coordinate, 156–159
Double curvature, law of, 142–147

Earth
    curvature of, 5, **6, 24**
    curved space around, **149**
    curved spacetime geometry, 124
    density of, 211
    gravitomagnetism, 232–233
    orbit, 171
    and planetary motion, 180
    Schwarzschild geometry, 124
    sphericity, **123**
Einstein, Albert, ix, 10–11
    conserved energy formula, 172
    on curvature of space, 5–7
    on expansion of universe, 230
    on free float, 11–14, **20,** 22
    general theory of relativity, 2, 3, 12–14, 124, 160, 171–172, 187, 230, 241, 243
    geometric account of planetary motion, 172
    geometrodynamic field equation, 118–120, 232
    on gravity, 11–15, 68
    on gravity radiation, 188
    on motion, 27
    on Newton, 2, 3
    on resistance and mass, 7
    on spacetime curvature, 29–30
    special theory of relativity, 7–9, 12–14, 30, 37
Einstein-Cartan equation, 118, **119**
Einstein-Friedmann model universe, 230
Einstein-Lense-Thirring frame dragging, 232, 233
Electric charge
    black holes, 218
    and electromagnetic waves, 191–195
Electric lines of force, 192–195
Electric moment, 199–200
Electricity and Matter (Thomson), 192
Electromagnetic induction, 8, 37
Electromagnetic radiation, 195

Hyperbola, 143, **144**
Hyperion (center of Earth), 58

In-the-surface curvature, 136–137, 140–142
Inertia
    in collisions, 95
    mass and motion, 26–27
    Newton on, 7
    and planetary motion, 178, 179
Interval, 35–53, 99
    null, 43, 44, 72
    spacelike, 36–37, 43, 44, 46
    and spacetime, 33, 161, 162
    timelike, 36–37, 41, 43, 44, 46
    zero, 43, 46, 51
Invariant
    distance, 33
Inverse-square law, 195
    electric charge, 190–192
    and gravity, 191
Io, 180
Irreducible mass of black holes, 218, 222

James, Henry, 190
Jupiter
    mass of, 180
    moons, **165, 180**

Kalamazoo, Michigan, 22, 32
Kalckar, Jørgen, 187–188
Kant, Immanuel, 236
KDKA broadcasting station
    (Pittsburgh), 191
Kelvin, Lord (William Thomson), 188–190, 220
Kepler, Johann, 89
    law of equal area in equal time, 176
    laws of planetary motion, 178–180
Kepler 1-2-3 law, 179–180, 236
Kinetic energy
    planetary motion, 170–171, 182
Kink
    electromagnetic waves, 190–195
    in worldlines, **49–51**
Klein, Felix, 33
Kokomo, Michigan, 22
Koran, 36
Krotkov, Robert, 27
Kyoto, Japan, 22, 32

Laboratory time, 30
LAGEOS I, 233
Laplace, Pierre-Simon, 210–211
Lapparent, V., 238, **239**
Laser beams
    gravity-wave detectors, 204
Leaning Tower of Pisa, 24, 25
Leibniz, Gottfried Wilhelm von, 121
    calculus, 154
    on space and time, 3
Lense, J., 232
Lensing, gravitational, 237–238
Library of Alexandria, **143, 144**
Light
    black holes, 210–211
    zero-interval links, **35**
Light, speed of, 31
    aging and spacetime, 49, 50
    free float, 37
    and interval, 40–43
    and space travel, 52
Linear momentum
    planetary motion, 169
Lines of force
    black holes, 225
    electric, 192–195
    gravity radiation, 196
Live (rotating) black hole, 213–218, 223
LMC X-1, 212
LMC X-3, 212
Lobachevsky, Nikolai Ivanovich, 4
Location
    and surface curvature, 140

Mach, Ernst, 7, 14, 232
Marconi, Guglielmo, 189
Mars
    motion of, 178
Marshall Islands, 104
Mass
    asymmetry of and gravity waves, 201–203
    attraction of, 62–63
    black holes, 210–213, 217–219, 222, 224
    center of attraction and planetary motion, 179–180
    curvature of Schwarzschild geometry, 137
    and energy, 9
    and energy loss, 187–188

geometric measure of, 87
geometrodynamic field equation, 119
gravitational attraction of, 210
gravitomagnetism, 232
and gravity waves, 186, 196–198
grip of spacetime on, 11, **13,** 72, 88, 94–107, 113, 120, 130, 196–198, 228, 243
and heat, 103
and inertia, 232
missing, 234–240
and momentum-energy (momenergy), 97–102, 104
and motion, 26–27
nonsphericity of, 200–201
Schwarzschild geometry, 124
spherical symmetry and gravity waves, 200–202
of universe, 240–243
and x-ray emission, 212
Mass, center of
    space geometry and spacetime geometry around, 149–163
Mass-free space
    curvature, 128–132
Mass-in-motion (*see* Momentum-energy)
Massachusetts Institute of Technology
    gravity-wave detector, 204
Mathematics
    Pythagorean Greeks, **144**
Maximal aging, principle of, 166–169, 174–175
Maxwell, James Clerk, 8
Measure of "creation," 105
Mechanics, science of, 3, 71–72, 116
Mediterranean Sea
    tides, 89
Mejia, Christopher A., 175
Menaechmus, **144**
Mercury
    orbit, 177–182
Metric, Schwarzschild space, 150–163
Michell, Reverend John, viii, 210–211
Mike nuclear explosion (1952), 103, 104
Milky Way, 18, 213
    energy, 224
    Galileo on, 235
Minkowski, Hermann, 30
Misner, Charles W., 179

Other books in the Scientific American Library Series

POWERS OF TEN
by Philip and Phylis Morrison and
the Office of Charles and Ray Eames

HUMAN DIVERSITY
by Richard Lewontin

THE DISCOVERY OF SUBATOMIC PARTICLES
by Steven Weinberg

THE SCIENCE OF MUSICAL SOUND
by John R. Pierce

FOSSILS AND THE HISTORY OF LIFE
by George Gaylord Simpson

THE SOLAR SYSTEM
by Roman Smoluchowski

ON SIZE AND LIFE
by Thomas A. McMahon and John Tyler Bonner

PERCEPTION
by Irvin Rock

CONSTRUCTING THE UNIVERSE
by David Layzer

THE SECOND LAW
by P. W. Atkins

THE LIVING CELL, VOLUMES I AND II
by Christian de Duve

MATHEMATICS AND OPTIMAL FORM
by Stefan Hildebrandt and Anthony Tromba

FIRE
by John W. Lyons

SUN AND EARTH
by Herbert Friedman

EINSTEIN'S LEGACY
by Julian Schwinger

ISLANDS
by H. William Menard

DRUGS AND THE BRAIN
by Solomon H. Snyder

THE TIMING OF BIOLOGICAL CLOCKS
by Arthur T. Winfree

EXTINCTION
by Steven M. Stanley

MOLECULES
by P. W. Atkins

EYE, BRAIN, AND VISION
by David H. Hubel

THE SCIENCE OF STRUCTURES AND
MATERIALS
by J. E. Gordon

SAND
by Raymond Siever

THE HONEY BEE
by James L. Gould and Carol Grant Gould

ANIMAL NAVIGATION
by Talbot H. Waterman

SLEEP
by J. Allan Hobson

FROM QUARKS TO THE COSMOS
by Leon M. Lederman and David N. Schramm

SEXUAL SELECTION
by James L. Gould and Carol Grant Gould

THE NEW ARCHAEOLOGY AND
THE ANCIENT MAYA
by Jeremy A. Sabloff